WAVE PROPAGATION
An Invariant Imbedding Approach

Mathematics and Its Applications

Richard Bellman

Late of the Department of Electrical Engineering, University of Southern California, Los Angeles, U.S.A.

and

Ramabhadra Vasudevan

The Institute of Mathematical Sciences, Madras, India

WAVE PROPAGATION

An Invariant Imbedding Approach

D. Reidel Publishing Company

A MEMBER OF THE KLUWER ACADEMIC PUBLISHERS GROUP

Dordrecht / Boston / Lancaster / Tokyo

Library of Congress Cataloging-in-Publication Data

CIP

Bellman, Richard Ernest, 1920–
Wave propagation.

(Mathematics and its applications)
Includes bibliographies and index.
1. Wave equation–Numerical solutions. 2. Invariant imbedding. 3. Dynamic programming. I. Vasudevan, Ramabhadra, 1926– . II. Title. III. Series.
QA927.B44 1986 530.1'24 85–31252
ISBN-13: 978-94-010-8811-4 e-ISBN-13: 978-94-009-5227-0
DOI: 10.1007/978-94-009-5227-0

Published by D. Reidel Publishing Company
P.O. Box 17, 3300 AA Dordrecht, Holland

Sold and distributed in the U.S.A. and Canada
by Kluwer Academic Publishers
190 Old Derby Street, Hingham, MA 02043, U.S.A.

In all other countries, sold and distributed
by Kluwer Academic Publishers Group,
P.O. Box 322, 3300 AH Dordrecht, Holland

TABLE OF CONTENTS

EDITOR'S PREFACE

Approach your problems from the right end and begin with the answers. Then one day, perhaps you will find the final question.

'The Hermit Clad in Crane Feathers' in R. van Gulik's *The Chinese Maze Murders*.

It isn't that they can't see the solution. It is that they can't see the problem.

G.K. Chesterton. *The Scandal of Father Brown* 'The point of a Pin'.

Growing specialization and diversification have brought a host of monographs and textbooks on increasingly specialized topics. However, the "tree" of knowledge of mathematics and related fields does not grow only by putting forth new branches. It also happens, quite often in fact, that branches which were thought to be completely disparate are suddenly seen to be related.

Further, the kind and level of sophistication of mathematics applied in various sciences has changed drastically in recent years: measure theory is used (non-trivially) in regional and theoretical economics; algebraic geometry interacts with physics; the Minkowsky lemma, coding theory and the structure of water meet one another in packing and covering theory; quantum fields, crystal defects and mathematical programming profit from homotopy theory; Lie algebras are relevant to filtering; and prediction and electrical engineering can use Stein spaces. And in addition to this there are such new emerging subdisciplines as "experimental mathematics", "CFD", "completely integrable systems", "chaos, synergetics and large-scale order", which are almost impossible to fit into the existing classification schemes. They draw upon widely different sections of mathematics. This programme, Mathematics and Its Applications, is devoted to new emerging (sub)disciplines and to such (new) interrelations as exempla gratia:

- a central concept which plays an important role in several different mathematical and/or scientific specialized areas;
- new applications of the results and ideas from one area of scientific endeavour into another;
- influences which the results, problems and concepts of one field of enquiry have and have had on the development of another.

The Mathematics and Its Applications programme tries to make available a careful selection of books which fit the philosophy outlined above. With such books, which are stimulating rather than definitive, intriguing rather than encyclopaedic, we hope to contribute something towards better communication among the practitioners in diversified fields.

Because of the wealth of scholarly research being undertaken in the Soviet Union, Eastern Europe, and Japan, it was decided to devote special attention to the work emanating from these particular regions. Thus it was decided to start three regional series under the umbrella of the main MIA programme.

Invariant imbedding can be regarded as, indeed is, a particularly successful instance of one of the most pervasive idea of modern mathematics, namely that it is often advantageous to study a problem not in isolation but as a member of a family of (similar) problems depending on an additional parameter. This additional parameter can be, for example, the size of the problem as in problems of transport theory (radiative transfer, transmission lines, Riccati equations) where the invariant imbedding method found its origin. This general idea, under the name of "the continuation method" was a favourite technique of Poincaré. It consists of imbedding a given problem in a one-parameter family of problems depending on an auxiliary parameter s and considering the solvability of the problem as s varies. Modern developments of this idea, highly successful both theoretically and numerically include of course bifurcation theory, the various homotopy methods of solving non-linear equations (also called continuation method), and most emphatically the subject of the present book: "invariant imbedding". Besides the numerical advantages such as efficiency, e.g. in the case of Riccati equations, there are theoretical advantages coming from the fact that one studies not only the problem itself but also (some of its) deformations/perturbations. And this in turn has practical value in terms of information concerning robustness of the model. More specifically the present volume uses invariant imbedding ideas and techniques (as well as the far from unrelated dynamic programming ideas) to deal with wave equations. Thus it offers an insightful novel approach, of considerable numerical power, relevant to such areas as quantum physics, acoustics, chemical physics, acoustic engineering and (of course) radiative transfer. Indeed a course on the first two of the subjects just named could well be based on this book.

The unreasonable effectiveness of mathematics in science ...

Eugene Wigner

Well, if you know of a better 'ole, go to it.

Bruce Bairnsfather

What is now proved was once only imagined.

William Blake

As long as algebra and geometry proceeded along separate paths, their advance was slow and their applications limited.

But when these sciences joined company they drew from each other fresh vitality and thenceforward marched on at a rapid pace towards perfection.

Joseph Louis Lagrange.

Bussum, February 1986

Michiel Hazewinkel

PREFACE

The purpose of this book is to apply invariant imbedding and dynamic programming to the wave equation.

In Chapter I we present a few of the methods currently used in the treatment of the wave equation. In Chapter II we give the celebrated WKB method. In Chapter III we present the fundamentals of invariant imbedding using a simple transport process as a model. In Chapter IV we give some applications of invariant imbedding to the wave equation. In Chapters V and VI we present the Bremmer Series and some generalizations. In Chapter VII we consider some time-dependent processes. In Chapter VIII we consider some asymptotic properties of the wave equation. In Chapter IX we consider some operator techniques which are useful in the study of the wave equation and other parts of mathematical physics. In Chapter X we present the common variational principles which are used. In studying variational principles, dynamic programming is useful. In Chapter XI we use dynamic programming for the solution of some wave equations. In Chapter XII we show how different approximations can be used to obtain approximate solutions of the wave equation.

The present volume does not exhaust all the methods. The Chapters therein will describe (1) the Feynman-Kac path integral formalism which has become an important tool even in field theory, (2) the inverse problems of determining the potentials, Gelfand-Levitan equations, etc., (3) using imbedding methods for non-linear equations deterministic as well as stochastic, (4) imbedding methods in transport theory with applications, (5) wave propagation in random media with applications in remote sensing, acoustic waves in the ocean, echo sounding and several other topics of current interest.

This work has been sponsored by the National Science Foundation, 'Invariant Imbedding in Engineering, Physiology, and Applied Science', headed by Dr Frederick W. Weingarten under Grant No. DCR 74-23539; and 'New Analytic and Computational Techniques for Ordinary and Partial Differential Equations of Mathematical Physics', headed by Dr Robert Agins under Grant No. MPS74-15650.

POSTSCRIPT

Professor R.E. Bellman, the senior author of this book, is no longer with us. He was my "Guru" in many ways and, as everybody knows, he had to go through a very difficult period during the last eleven years of his life due to serious illness. With the help of his devoted wife Mrs. Nina Bellman, he endured this with undaunted and steadfast courage and accomplished much creative work.

Richard Bellman was a master of the spoken and the written word, and he contributed profoundly and in great measure to pure and applied mathematics all through his life. With his steadfast conviction that mathematics should not be an ivory tower but should be linked with everyday life, he opened up new vistas of thought, in a wide range of subjects: dynamic programming, stability theory, imbedding methods, multistage decision process, stochastic control theory, etc. He pioneered new ideas like expert systems, artificial intelligence, biomedical engineering and control, etc. His ideas are being used daily in areas covering a wide spectrum of the physical sciences, control engineering, management sciences, medicine, psychology, etc.

I feel very proud and honoured to have been associated with Professor Bellman, whose work has made such a significant contribution to our understanding in many ways.

His many friends and colleagues were not only struck by his brilliance and humor and his determination under difficult circumstances, but also by his lavish humanity which made him give and give joyfully to all those who needed his help in different ways.

I thank profoundly his devoted wife Mrs. Nina Bellman who made effective arrangements and facilities for the preparation of this book and also Mrs. Jean Britton, his secretary, who typed the manuscript, and my colleagues for checking the proofs. As Professor Bellman used to say "A

book never gets completed, it escapes the authors ".
I also thank the publishers for their help and kind patience.

March 1986 R. VASUDEVAN

CHAPTER I

INTRODUCTION

1. INTRODUCTION

In this chapter we will briefly present a survey of some of the well-known wave equations, describing waves traveling along inhomogeneous media and the attempts to arrive at solutions under different approximations. The electromagnetic fields and the wave propagation associated with them have been studied in great detail. These are vector wave equations for the electric $\vec{E}$ and magnetic $\vec{H}$ vector fields [1]. The scalar Schrödinger wave equation of non-relativistic quantum mechanics has been the subject of intense analysis [2]. The study of the passage of an electromagnetic wave through an inhomogeneous medium is vital to the great advances made in optics and hence we introduce below the ideas relating to the passage of a plane wave through a stratified medium, the electric and magnetic fields being perpendicular to each other in a plane perpendicular to the direction of the wave. They can be shown to satisfy linear equations provided the dielectric constant ε and magnetic permeability μ are independent of the applied fields. The case in which these quantities depend on the applied fields leads to nonlinear equations, which can be analyzed by the imbedding techniques presented in this monograph. We will deal with mainly linear wave propagation in this volume. The solutions are subject to certain conservation laws, and the concepts of reflectivity R and transmittivity T play important roles. In investigating the optics of thin films the concept of the stratified medium refractive indices constant in each layer but varying in different layers provides the key to the analysis of the problem [3]. We shall see later that this corresponds to the 'principle of localization' which leads to a simplified and intuitive picture of wave propagation in inhomogeneous media, to be studied utilizing the 'principle of invariant imbedding' in this volume [4]. Following these ideas much work has been done on seismic waves, elastic waves, acoustic waves, electromagnetic waves, etc., and also in connection with the reflection of radio waves in the earth's atmosphere [5].

Similar analogies carry over to the field of electrical transmission also [6], [7], [8].

2. PROPAGATION OF A PLANE ELECTROMAGNETIC WAVE IN A STRATIFIED MEDIUM

The theory of stratified media is of considerable importance in optics and other fields in connection with multilayers, i.e., a succession of thin plane parallel films. Inside each layer the optical properties pertaining to propagation may be considered constant throughout each plane perpendicular to a given direction, say the Z direction of the cartesian coordinate system. The dielectric constant ε and magnetic permeability μ are functions of Z only. We consider a plane parallel time harmonic wave through these layers.

When a plane wave falls on a boundary between two media of different optical properties, it is split into a transmitted wave proceding into the second medium and a reflected wave propagated back into the first medium. The plane specified by the direction of the incident wave and normal to the boundary at the point of incidence is the plane of incidence and Snell's law connects the angles of incidence θ_i and transmission θ_t with the refractive indices of the two media.

$$\frac{\mathrm{Sin}\ \theta_i}{\mathrm{Sin}\ \theta_t} = \frac{\sqrt{\varepsilon_2\mu_2}}{\sqrt{\varepsilon_1\mu_1}} = \frac{n_2}{n_1} = n_{12}\ , \tag{1}$$

where n_1 and n_2 are the refractive indices of the first and second medium respectively referred to vacuum and n_{12} their relative refractive index. The dielectric constants and magnetic permeability of the two media are ε_1 and ε_2; and μ_1 and μ_2 respectively. The components of the electric vector of the incident field can be split up into $E^i_{\parallel}$ and $E^i_{\perp}$ being parallel and perpendicular to the incident plane. Hence, the X, Y and Z components of the incident $\vec{E}$ field and similarly those of the reflected and transmitted fields can be obtained knowing θ_i, θ_r and θ_t, the angles of incidence, reflection* and the transmission respectively at the boundary. Once the E vector is known, the magnetic vector is given by

---

* In the current notation $\theta_r = (\pi-\theta_i)$

$$\vec{H} = \sqrt{\frac{\varepsilon}{\mu}}\,\vec{s} \times \vec{E} \,, \tag{2}$$

where $\vec{s}$ is the unit vector in the direction of propagation. The boundary conditions at the surface of separation require that the tangential component of E as well as H be continuous. Hence, we arrive at the following relations* for $T_{\|}$, $T_{\perp}$ the transmitted amplitudes parallel and perpendicular to the incident plane and $R_{\|}$ and $R_{\perp}$, the analogous reflected amplitudes in relation to the incident parallel and perpendicular amplitudes $E^{i}_{\|}$ and $E^{i}_{\perp}$**

$$T_{\|} = \frac{2n_1 \cos\theta_i}{n_2 \cos\theta_i + n_1 \cos\theta_t} E_{\|} \tag{3}$$

$$T_{\perp} = \frac{2n_1 \cos\theta_i}{n_1 \cos\theta_i + n_2 \cos\theta_t} E_{\perp} \tag{4}$$

$$R_{\|} = \frac{n_2 \cos\theta_i - n_1 \cos\theta_t}{n_2 \cos\theta_i + n_1 \cos\theta_t} E_{\|} \tag{5}$$

$$R_{\perp} = \frac{n_1 \cos\theta_i - n_2 \cos\theta_t}{n_1 \cos\theta_i + n_2 \cos\theta_t} E_{\perp} \tag{6}$$

These are the well-known Fresnels' formulae [1].

The energy of the incident beam is divided into the reflected and transmitted components and the conservation of energy principle applies at each boundary. The light intensity is the modulus of the Poynting vector, and is given by

$$S = \frac{cn}{4\pi} E^2 \,, \tag{7}$$

where 'n' is the refractive index of the mdeium and 'c' the velocity of light in vacuum.

The energy incident per unit area at the boundary per unit time is given by

---

* In what follows we assume $\mu = 1$

** We follow the method in reference [1] in these discussions

$$S^i = \frac{cn_1 |E^i|^2}{4\pi} \cos \theta_i \tag{8}$$

and the energy of the reflected and transmitted components are

$$S^r = \frac{cn_1}{4\pi} |R|^2 \cos \theta_i \tag{9}$$

$$S^t = \frac{cn_2}{4\pi} |T|^2 \cos \theta_t . \tag{10}$$

The ratios $\bar{R}$ and $\bar{T}$ are defined as

$$\bar{R} = \frac{S^r}{S^i} = \frac{|R|^2}{|E^i|^2} \; ; \quad \bar{T} = \frac{S^t}{S^i} = \frac{|T|^2}{|E^i|^2} \frac{n_2 \cos \theta_t}{n_1 \cos \theta_i} . \tag{11}$$

They are called the reflectivity and transmittivity coefficients. Utilizing the expressions (3) through (6) we find

$$\bar{R} + \bar{T} = 1 \tag{12}$$

which is an expression of conservation of energy. To arrive at (12), it would be easier if Equations (3) through (6) are reexpressed as

$$T_{\parallel} = \frac{2 \sin \theta_t \cos \theta_i}{\sin (\theta_i + \theta_t) \cos (\theta_i - \theta_t)} E_{\parallel} \tag{3a}$$

$$T_{\perp} = \frac{2 \sin \theta_t \cos \theta_i}{\sin (\theta_i + \theta_t)} E_{\perp} \tag{4a}$$

$$R_{\parallel} = \frac{\tan (\theta_i - \theta_t)}{\tan (\theta_i + \theta_t)} E_{\parallel} \tag{5a}$$

$$R_{\perp} = \frac{- \sin (\theta_i - \theta_t)}{\sin (\theta_i + \theta_t)} E_{\perp} \tag{6a}$$

and $\bar{R}_{\parallel} + \bar{T}_{\parallel}$ and $\bar{R}_{\perp} + \bar{T}_{\perp}$ are each equal to unity.

3. BASIC DIFFERENTIAL EQUATIONS OF THE ELECTROMAGNETIC FIELD*

As is well known, the following are Maxwell's equations of an electromagnetic field together with the constitutive relations

$$\text{Curl } H = \frac{1}{c}\dot{D} + \frac{4\pi}{c} J \tag{13}$$

$$\text{Curl } E + \frac{1}{c}\dot{B} = 0 \tag{14}$$

$$\text{Div } D = 4\pi\rho \tag{15}$$

$$\text{Div } B = 0 \tag{16}$$

and

$$J = \sigma E, \quad D = \varepsilon E, \quad \text{and} \quad B = \mu H, \tag{17}$$

J is the electric current, D the displacement, B the magnetic induction, and the dot represents differentiation with respect to time. For a dielectric $\sigma = 0$ the above equations are to be solved for a given distribution of charges and currents.

Consider a plane, time harmonic electromagnetic wave propagated through a stratified medium. In the case when the wave is linearly polarized with its electric vector perpendicular to the place of incidence, we call it Transverse Electric wave (TE) and when its magnetic vector is perpendicular to the plane of incidence we call it the Transverse Magnetic wave (TM). An arbitrarily polarized wave can be resolved into two waves one of which is a TE wave and the other a TM wave which are mutually independent.

Focussing our attention on Maxwell's equation propagating in a medium free of charges or currents, we have from Equations (13) and (14)

$$\frac{1}{\mu}(\text{Curl Curl } E) + (\text{Grad } \frac{1}{\mu}) \times (\text{Curl } E) + \frac{\varepsilon}{c^2}\ddot{E} = 0 \tag{18}$$

and from (15) we obtain

$$\nabla^2 E + (\text{grad log } \mu) \times (\text{Curl } E) + \text{grad}(E \cdot \text{grad log } \varepsilon) = \frac{\mu\varepsilon}{c^2}\ddot{E}. \tag{19}$$

If we take the plane of incidence to be the Y - Z plane,

---

* Herein E, H, B, D and J are to be taken appropriately as vector quantities of the E.M. field.

considering the TE wave propagating along the Z axis, as a plane harmonic wave of frequency ω, we arrive at*

$$\frac{\partial^2 E_x}{\partial y^2} + \frac{\partial^2 E_x}{\partial z^2} + \frac{\mu\varepsilon}{c^2}\omega^2 E_x = \left(\frac{\partial \log \mu}{\partial z}\right)\frac{\partial E_x}{\partial z} . \qquad (20)$$

Remembering the fact taht $E_z = E_y = 0$, we know from Maxwell's equations that H_y, H_z and E_x are only functions of y and z. Equation (20) is the x component of the wave equation (19). We can solve (20) by separation of variables assuming that

$$E_x(y, z) = Y(y)U(z). \qquad (21)$$

Consequently, we have

$$E_x = U(z)e^{i(k_0 ay - \omega t)} , \qquad (22)$$

where U(z) and Y(y) satisfy the equations

$$\frac{1}{Y}\frac{d^2Y}{dy^2} = -k_0^2 a^2; \quad k_0 = \omega/c \qquad (23)$$

and

$$\frac{d^2U}{dz^2} - \frac{d(\log \mu)}{dz}\frac{dU}{dz} + (n^2 - a^2)k_0^2 U = 0 \qquad (24)$$

$$n^2 = \varepsilon\mu$$

A wave equation similar to Equation (19) can be obtained for H field and hence H_y and H_z can be obtained as

$$H_y = V(z)e^{i(k_0 ay - \omega t)} \qquad (25)$$

$$H_z = W(z)e^{i(k_0 ay - \omega t)} . \qquad (26)$$

Maxwell's equations in the component form for TE waves are

---

* We assume ε and μ vary only along the Z-direction.

$$\left.\begin{aligned}
&\frac{\partial H_z}{\partial y} - \frac{\partial H_y}{\partial z} + \frac{i\varepsilon\omega}{c} E_x = 0 \\
&\frac{\partial H_x}{\partial z} - \frac{\partial H_z}{\partial x} = 0 \\
&\frac{\partial H_y}{\partial x} - \frac{\partial H_x}{\partial y} = 0
\end{aligned}\right\} \tag{27}$$

$$\left.\begin{aligned}
&\frac{i\omega\mu}{c} H_x = 0 \\
&\frac{\partial E_x}{\partial z} - \frac{i\omega\mu}{c} H_y = 0 \\
&\frac{\partial E_x}{\partial y} + \frac{i\omega\mu}{c} H_z = 0
\end{aligned}\right\} . \tag{28}$$

From the above two sets, we can obtain the relations for U, V and W as

$$\frac{dV}{dZ} = i\, k_0 (aW + \varepsilon U) \tag{29}$$

$$\frac{dU}{dZ} = i\, k_0\, \mu V \tag{30}$$

$$aU + \mu W = 0. \tag{31}$$

Also by elimination from the above equations, we obtain the second order equation for U and V given by the following which are the same as those in Equation (24)

$$\frac{d^2U}{dZ^2} - \frac{d \log(\mu)}{dZ} \frac{dU}{dZ} + k_0^2 (n^2 - a^2) U = 0 \tag{32}$$

$$\frac{d^2V}{dZ^2} - \frac{d \log(\varepsilon - a^2/\mu)}{dZ} \frac{dV}{dZ} + k_0^2 (n^2 - a^2) V = 0. \tag{33}$$

Thus in the case of media described above, for the TE wave, the E and H field can be calculated by solving the two second order linear equations (32) and (33). By the symmetry of Maxwells' equations a similar set of equations for the TM wave can be obtained simply by interchanging μ and ε as well as E and -H simultaneously. For the present set from Equation (22) the E_x field has a modulus $|U(z)|$ and might have a phase φ. The total phase $i(t - (\vec{r}.\vec{s})/v)\omega$ will be of the form $-i\{\omega y[(\sin \theta_i)/v] + \omega z[(\cos \theta_i)/v] - \omega t\}$ where v is the velocity of the medium and y-z the incident plane, θ_i the incident angle. Hence the coefficient of y in the phase at the point of incidence will be $(\omega \sin \theta_i)/v = k_0 n \sin \theta_i$. Hence 'a' in the above equations (22, 25, 26) is given by $n \sin \theta$, (where n is the refractive index of the medium) and this will remain a constant during the propagation and constitutes a generalization of Snell's law.

In fact solving either for U or V is adequate to obtain the complete knowledge of the fields in the TE case in view of the equations, (29) to (31). If in the first order Equations (29) and (30) the factor 'a' is taken to be zero, these are analogous to the equations appearing in electrical transmission lines thoery [6]. Thus the study of the propagation of a vector wave like a plane harmonic electromagnetic wave, through a stratified medium, can be related to solutions of second order differential equations of quantities like U and V. Since the inhomogeneity of the medium is only along one direction, U and V are not coupled and we have only one independent variable. To solve these, many approximations, guided by intuition and also by the physics of the problem, are in use as we describe in the next section.

4. PROPAGATION OF E.M. WAVES THROUGH MULTILAYERS

Since U(z) and V(z) satisfy second order differential equations (32) and (33), they may be expressed as linear combinations of two independent sets of solutions (U_1, U_2) and (V_1, V_2). Also from the first order equation (29), (30) and (31), we can show that the determinant

$$D = \begin{vmatrix} U_1 & V_1 \\ U_2 & V_2 \end{vmatrix} \tag{34}$$

is an invariant of the system.

The most convenient choice of the particular solution is

$$\begin{aligned} U_1 &= f(z); \qquad U_2 = F(z) \\ V_1 &= g(z); \qquad V_2 = G(z) \end{aligned} \tag{35}$$

such that

$$f(0) = G(0) = 0 \text{ and } F(0) = g(0) = 1. \tag{36}$$

Hence, if U_0 and V_0 are the initial values for U and V, we have for any Z plane

$$\vec{Q} = \begin{bmatrix} U(z) \\ V(z) \end{bmatrix} = \begin{bmatrix} F(z) & f(z) \\ G(z) & g(z) \end{bmatrix} \begin{bmatrix} U_0 \\ V_0 \end{bmatrix} = [N]\vec{Q}_0. \tag{37}$$

Inverting the above relation we can also write

$$\vec{Q}_0 = [N]^{-1}\vec{Q} = \begin{bmatrix} g(z) & -f(z) \\ -G(z) & F(z) \end{bmatrix} \vec{Q}(z) \tag{38}$$

since N is unimodular.

In any particular filament the ε and μ can be taken to be constant and hence U and V satisfy in this homogeneous filament wave equations (see Equations (32) and (33) and the definition of a).

$$\frac{d^2U}{dZ^2} + k_0^2\, n^2 \cos^2\theta\, U = 0 \tag{39}$$

$$\frac{d^2V}{dZ^2} + k_0^2\, n^2 \cos^2\theta\, V = 0. \tag{40}$$

Hence U(z) can be written as

$$U(z) = A \cos\,(k_0 nz \cos\,\theta) + B \sin\,(k_0 nz \cos\,\theta) \tag{41}$$

and in view of Equations (29) to (31) we have

$$V(z) = \frac{1}{i}\sqrt{\frac{\varepsilon}{\mu}} \cos \theta[B \cos (k_0 nz \cos \theta) - A \sin (k_0 nz \cos \theta)] \tag{42}$$

We can therefore obtain $M = N^{-1}$ from the particular solutions of U and V with the required initial conditions on them as

$$M(z) = \begin{bmatrix} \cos (k_0 nz \cos \theta) & -\frac{i}{p} \sin (k_0 nz \cos \theta) \\ -ip \sin (k_0 nz \cos \theta) & \cos (k_0 nz \cos \theta) \end{bmatrix} \tag{43}$$

with

$$p = \sqrt{\frac{\varepsilon}{\mu}} \cos \theta. \tag{44}$$

For a series of multilayers arranged as a pack, the initial vector at $z = 0$ is related to the vector at $z = z_n$ by

$$\vec{Q}_0 = M_1(z_1)M_2(z_2 - z_1)M_3(z_3 - z_2) \ldots M_n(z_n - z_{n-1})\vec{Q}(z_n), \tag{45}$$

where $M_i(z_i - z_{i-1})$ are transfer matrices which are unimodular. The constancy of the determinant of M implies conservation of energy. In the case of a TM wave p should be replaced by q given by

$$q = \sqrt{\frac{\mu}{\varepsilon}} \cos \theta. \tag{46}$$

If the medium is considered as a pack or pile of N thin films of thickness δz_i which is very small, we can write

$$M = \prod_{j=1}^{N} M_j = \begin{bmatrix} 1 & -ik_0 B' \\ -ik_0 A' & 1 \end{bmatrix}, \tag{47}$$

where

$$A' = \sum_j p_j n_j \delta z_j \cos \theta_j = \sum_j \left(\varepsilon_j - \frac{a^2}{\mu_j}\right) \delta z_j \tag{48}$$

and

$$B' = \sum_j \frac{n_j}{p_j}\, \delta z_j \cos\theta_j = \sum_{j=1}^{N} \mu_j\, \delta z_j. \tag{49}$$

In the limit as N tends to ∞ A' and B' can be replaced by appropriate integrals. Hence, in the continuous limit we have

$$M = \begin{bmatrix} 1 & -ik_0\int \mu\, dz \\ -ik_0\int\left(\varepsilon - \frac{a^2}{\mu}\right) dz & 1 \end{bmatrix} \tag{50}$$

Let us now turn to a homogeneous slab of thickness h of constant values ε_2 and μ_2 placed in between media 1 and 3 having constants (ε_1, μ_1) and (ε_3, μ_3) respectively, and semi-infinite in extension on both sides. Let us call the mid slab medium 2. Let a plane wave fall from medium 1 onto the slab of medium 2 on the surface $z = 0$, and emerge $z = z_3$ into medium 3. Let us call E^i an incident amplitude at $z = 0$ and R the reflected amplitude in the medium 1. Similarly, let T be the transmitted amplitude in the medium at $z = z_3$. For a given slab of thickness h bounded by media 1 and 3 with constants ε_1, ε_3 and μ_1, μ_3 we have because of continuity conditions at separation of media 1 and 2

$$\begin{aligned} U_0 &= E^i + R \\ V_0 &= p_1(E^i - R) \\ U(z_3) &= T \quad \text{and} \\ V(z_3) &= p_3 T \end{aligned} \tag{51}$$

where R and T are reflected and transmitted amplitudes respectively. Hence, we have

$$\begin{bmatrix} E^i + R \\ p_1(E^i - R) \end{bmatrix} = \begin{bmatrix} M(h) \end{bmatrix} \begin{bmatrix} T \\ p_3 T \end{bmatrix}; \quad M = \begin{bmatrix} m_{11}(h) & m_{12}(h) \\ m_{21}(h) & m_{22}(h) \end{bmatrix}, \tag{52}$$

where M(h) is the transfer matrix for the homogeneous medium (43). From (52) we can calculate the reflection and transmission coefficients 'r' and 't' of the medium as shown below.

$$r = \frac{R}{E^i} = \frac{(m_{11} + m_{12}p_3)p_1 - (m_{21} + m_{22}p_3)}{(m_{11} + m_{12}p_3)p_1 + (m_{21} + m_{22}p_3)} \tag{53}$$

$$t = \frac{T}{E^i} = \frac{2p_1}{(m_{11} + m_{12}p_3)p_1 + (m_{21} + m_{22}p_3)} \tag{54}$$

We also know that

$$m_{11} = m_{22} = \cos(k_0 h n_2 \cos\theta_2); \; m_{12} = -\frac{i}{p_2}\sin(k_0 h n_2 \cos\theta_2)$$
$$m_{21} = -ip_2 \sin(k_0 h n_2 \cos\theta_2); \quad p_j = n_j \cos\theta_j \quad j=1,\,2,\,3 \tag{55}$$

where θ_1 and θ_2 are angles of incidence and refraction at the face $z = 0$, between media 1 and 2. The Fresnel's formulae for reflection and transmission coefficients at each boundary, i.e., at the boundaries between media 1 and 2 and media 2 and 3 are given according to Equation (4) and (6) for a TE wave. We have, assuming $\mu = 1$ for all media,

$$r_{12} = \frac{p_1 - p_2}{p_1 + p_2} \tag{56}$$

$$t_{12} = \frac{2p_1}{p_1 + p_2} \tag{57}$$

Similar quantities r_{23} and t_{23} can be written down. Hence, for the total medium of thickness h, the reflection and transmission coefficients (53) and (54) are given by

$$r = \frac{r_{12} + r_{23}\, e^{2i\beta}}{1 + r_{12}r_{23}e^{2i\beta}} \tag{58}$$

$$t = \frac{t_{12}t_{23}e^{i\beta}}{1 + r_{12}r_{23}e^{2i\beta}} \tag{59}$$

where $\beta = k_0 h n_2 \cos \theta_2$.

For a multilayer the total reflection can be found by iteration of this result for a pile of such slabs. Thus, we have demonstrated the methods of obtaining the solutions of the field inside the medium given the initial conditions. Of course, even for the propagation of the electromagnetic wave which is a vector wave with coupling between the electric and magnetic fields we arrived at one dimensional wave equations for the simplified model consisting of layers of thin films [7] whose optical properties varied only along the z-axis of the medium. In the succeeding chapters we will see how first order nonlinear equations can be written down for the emerging fields generalizing the idea of the thin films; i.e., introducing the concept of localization [4] and employing the principle of imbedding methods.

Seismic Waves

The passage of seismic waves through strata of earth has been studied intensely in the context of exploration of oil and natural gas. Seismic prospecting is conducted either on land or at sea. Partial reflection and transmission of a seismic wave occurs at an interface where there is a change in acoustic impedence which is defined as the product of the velocity and the density of the rock. These waves are created by earthquakes or nuclear explosions. In exploration these are produced by chemical explosions or vibratory impacts of short durations. The sedimentary layers of the earth are of large dimensions and correspond to layered media and reflection and transmission occurs at the interfaces of these layers. The signal frequency is in the range of 20 to 60 MHz. Taking the transform of the time variable, the reflection and transmission functions for N layers can be obtained as was done in the case of the Stokes formula for a pile of layers, like the Equation (45). From the reflected signals one obtains a clue to the nature of the layers inside the earth's crust and a vast amount of work has been done in this field using the apparatus of a digital signal processing and time series analysis. The reader is referred to detailed work on this subject in references [13, 14 and 15].

5. THE SCHRÖDINGER EQUATION

Let us turn our attention to the familiar one dimensional scalar wave equation of quantum mechanics. The WKB solutions [9] of the Schrödinger equation have been of great importance in describing the quantum dynamics in a number of situations. The equation itself describes the wave function Ψ of a particle moving in a potential field as:

$$\frac{d^2\Psi}{dx^2} + \frac{2m(E - V)\Psi}{\hbar^2} = 0 \tag{60}$$

m is the mass, $2\pi\hbar$ the Planck's constant, E the energy of the particle and V(x) the potential in general a function of the variable x. The momentum of the particle p is given by

$$p = \sqrt{2m(E - V)} = \hbar k \tag{61}$$

and hence if V is constant the solution can, in general, be written as

$$\Psi = Ae^{ikx} + Be^{-ikx} \tag{62}$$

and k is real if $E \geq V$.

We examine the quantum mechanics of a particle incident from left with an energy E in a potential region at x = 0, such that E > V; V being constant. In region 1 its momentum $p_1 = \hbar k_1 = \sqrt{2mE}$ and in region 2 its momentum is $p_2 = \hbar k_2 = \sqrt{2m(E - V)}$. Then we have for a steady wave (or steady beam of particles) the time independent Schrödinger Equation (60) having solutions

$$\Psi(x) = Ae^{ik_1x} + Be^{-ik_1x} \qquad x < 0 \tag{63}$$

$$\Psi(x) = Ce^{ik_2x} \qquad x > 0 \tag{64}$$

The smooth joining of the wave functions at the point of separation or at the boundary of the two regions require Ψ and Ψ' to be continuous at x = 0. Hence, we have

$$A + B = C \tag{65}$$

$$k_1(A - B) = k_2C. \tag{66}$$

We solve for the ratios B/A and C/A the coefficients of reflection r, and transmission t at the boundary to be:

$$r = B/A = \frac{k_1 - k_2}{k_1 + k_2} \; ; \qquad t = C/A = \frac{2k_1}{k_1 + k_2} \; . \tag{67}$$

The current density of the steady beam is constant at every point and it is expressed as

$$j = \frac{\hbar}{2mi}\left(\Psi^* \frac{d\Psi}{dx} - \frac{d\Psi^*}{dx}\Psi\right). \tag{68}$$

Therefore j(x) in the two regions are

$$j = \frac{\hbar k_1}{m}\left(|A|^2 - |B|^2\right) \quad x < 0 \tag{69}$$

$$j = \frac{\hbar k_2}{m}|C|^2 \qquad x > 0 \tag{70}$$

The equality of the j values in (69) and (70) is assured by Equation (67). This conservation law is analogous to the energy conservation described in the case of optics. If we define $\bar{R} = r^2$ and $\bar{T} = k_2 t^2 k_1$ as reflectivity and transmitivity, we see that

$$\bar{R} + \bar{T} = 1 \tag{71}$$

as in the case of optics (see Equations (11) and (12)).

6. THE RECTANGULAR POTENTIAL BARRIER

In quantum mechanics if a particle with an energy E falls on a region of constant potential V extending from +a to −a, it emerges at the other side with changes in its phases and also gives rise to a reflected part of the wave behind the potential region. Classically, if incident energy E < V the particle does not penetrate. However, quantum mechanically the probability amplitude of the particle described by its wave function leaks out to the other side of the barrier unless V becomes infinite. Hence, the importance of such problems. Let us consider the three regions: $x < a$; $-a \leq x \leq a$ and $x > a$ and write down the wave functions assuming $E > V$.

$$\Psi(x) = \begin{cases} Ae^{ik_1x} + Be^{-ik_1x} & x < -a \\ Ce^{ik_2x} + De^{-ik_2x} & -a \le x \le a \\ Fe^{ik_1x} + Ge^{-ik_1x} & x > a \end{cases} \tag{72}$$

The boundary conditions at x = -a yields,

$$Ae^{-ik_1a} + Be^{ik_1a} = Ce^{-ik_2a} + De^{ik_2a} \tag{73}$$

$$Ae^{-ik_1a} - Be^{ik_1a} = \frac{k_2}{k_1}(Ce^{-ik_2a} - De^{ik_2a}). \tag{74}$$

This means

$$\begin{pmatrix} A \\ B \end{pmatrix} = [M_1]\begin{pmatrix} C \\ D \end{pmatrix} , \tag{75}$$

where the elements of the matrix M_1 are obtained from Equations (73) and (74). Similar boundary conditions at the end of the potential region yield another relation for the amplitudes at this end as

$$\begin{pmatrix} C \\ D \end{pmatrix} = [M_2]\begin{pmatrix} F \\ G \end{pmatrix} . \tag{76}$$

Hence, we have

$$\begin{pmatrix} A \\ B \end{pmatrix} = [M]\begin{pmatrix} F \\ G \end{pmatrix} ; \quad M_1M_2 = M, \tag{77}$$

where

$$[M] = \begin{bmatrix} m_{11} & m_{12} \\ m_{21} & m_{22} \end{bmatrix}$$

$$= \begin{bmatrix} (\cos 2k_2a - \frac{is_2}{2}\sin 2k_2a)e^{2ik_1a} & \frac{it_1}{2}\sin 2k_2a \\ \frac{-it_1}{2}\sin 2k_2a & e^{-2ik_1a}[\cos 2k_2a + \frac{i}{2}s_2 \sin 2k_2a] \end{bmatrix}$$

with

$$s_2 = \frac{k_2}{k_1} + \frac{k_1}{k_2} \;; \quad \text{and} \quad t_1 = \left(\frac{k_1}{k_2} - \frac{k_2}{k_1}\right). \tag{78}$$

It should be noted that M is unimodular and $m_{11} = m_{22}^*$; $m_{21} = m_{12}^*$. These can be related to the conservation of probability and time reversal invariance of the system. Computing the conservation of current we have

$$A^2 - B^2 = F^2 - G^2. \tag{79}$$

If G = 0 we have $\bar{R} + \bar{T} = 1$ as before.

7. THE JWKB SOLUTIONS

If the potential V(x) is not a constant but a complicated function of x, the Schrödinger equation has no simple analytic solution in general. However, under certain conditions very useful approximations can be found apart from other methods. The JWKB solution [10] or phase integral solutions [11] have been widely used. The original papers of Wentzel, Kramers, Brillouin and Jeffreys as well as those of Liouville and Green have been listed in [11]. If V is a slowly varying function of x, we can try a solution of the form

$$\Psi = e^{iy(x)} \;, \tag{80}$$

where y(x) the phase of the wave can be found to satisfy the equation

$$i \frac{d^2y}{dx^2} - \left(\frac{dy}{dx}\right)^2 + [k(x)]^2 = 0 \tag{81}$$

with

$$k(x) = \left[\frac{2m}{\hbar^2}(E - V)\right]^{\frac{1}{2}} \quad \text{for } E > V \tag{82}$$

The nonlinear Equation (81) can be solved iteratively if d^2y/dx^2 is quite small, and it can be assumed to be so if V(x) varies very slowly with x. The crudest approximation y_0 to y is given by

$$y_0(x) = \pm \int^x k(x)\, dx + C, \tag{83}$$

Feeding this into Equation (81),

$$\left[\frac{dy}{dx}\right]^2 = [k(x)]^2 + i\,\frac{d^2y}{dx^2} , \tag{84}$$

we can find y_1, y_2, etc., iteratively yielding y_{n+1} in terms of y_n where

$$y_{n+1}(x) = \pm \int^x \sqrt{k^2(x) + iy_n''(x)}\, dx + C_1 \tag{85}$$

for n = 0, we have

$$y_1 = \pm \int^x \sqrt{k^2(x) + ik'(x)}\, dx + C_1 \tag{86}$$

and if $|k'(x)| \ll |k^2(x)|$; we have the solution (87a)

$$y_1 \simeq \pm \int^x k(x)\, dx + \frac{i}{2} \log\,(k(x)) + C_1 \tag{87b}$$

and the wave function Ψ itself is given by

$$\Psi \simeq \frac{1}{\sqrt{k(x)}} \exp \pm i \int^x k(x)\, dx. \tag{88}$$

Since the wavelenght λ associated with the particle is $\lambda(x) = 2\pi/k(x)$, the condition (87a) expresses the fact that the change in momentum p ($= \hbar k$) of the particle over a wave-length is very much less than the momentum itself at the point x. That is

$$|\lambda\frac{dp}{dx}| \ll p(x). \tag{89}$$

This approximation breaks down when k(x) changes rapidly or when k vanishes, i.e., when E = V. This happens at the classical turning points. Near these regions more accurate solutions have to be used and different types of connection formulae have to be invented [11] to continue the solution from one region to the other. We will not go into these aspects now.

In many control theory problems and other situations we want to minimize a functional of the form

$$J(u) = \int_0^T [u'^2 + g(t)u^2]\, dt \tag{90}$$

under various restrictions. This leads to second order equations of the type mentioned earlier

$$u'' - g(t)u = 0 \tag{91}$$

with different types of boundary conditions.

Exploiting the physical picture of the process involved and employing the imbedding techniques, solutions, numerical as well as analytical, can be arrived at under different orders of approximations. The JWKB solution mentioned above is one such approximation. We shall talk about solutions of higher orders of approximations in later chapters, for the wave equations, describing the steady state as well as time dependent cases.

REFERENCES

[1] Born, Max, and E. Wolf Principles of Optics, Pergamon Press, London, 1964.
[2] Merzbacher, E., Quantum Mechanics, John Wiley & Sons, Inc., New York, 1962.
[3] Vasicek, A., Optics of Thin Films, North Holland Publishing Company, Amsterdam, 1950.
[4] Bellman, R., and R. Kalaba, 'Functional Equations, Wave Propagation and Invariant Imbedding', Journal of Mathematics and Mechanics 8 (1959), 683-704.
[5] Budden, K.G., Radio Waves in the Ionosphere, Cambridge Methuen Monographs, Methuen, London, 1981.
[6] Brekhovskikh, L.M., Waves in a Layered Media, Academic Press, Inc., New York, 1960.
[7] Schelkunoff, S.A., Electromagnetic Waves, D. Van Nostrand Co., Inc., New York, 1943.
[8] Abeles, F., Ann. de Physique 5 (1950), 596.
[9] Froman, Nanny and Per Olaf, JWKB Approximation - Contributions to the Theory, North Holland Publishing Co., New York, 1965.

[10] Kemble, E.C., <u>The Fundamental Principles of Quantum Mechanics</u>, Dover Publications, Inc., New York, 1958.

[11] Heading, J., <u>An Introduction to Phase Integral Methods</u>, Methuen and Co., Ltd., London, 1961.

[12] Heavens, O.S., <u>Optical Properties of Thin Solid Films</u>, Butterworth Scientific Pub., London, 1956.

[13] Robinson, E.A., <u>Multichannel Time Series Analysis with Digital Processing</u>, Holdenday, San Francisco, 1967.

[14] Markus, B., <u>Mathematical Aspects of Seismology</u>, Elsevier, Amsterdam, 1968.

[15] Robinson, E.A., and M.T. Silvia, <u>Digital Signal Processing and Time Series Analysis</u>, Pilot Edition, Holdenday, San Francisco, 1978.

CHAPTER II

EIKONAL EQUATION AND THE WKB APPROXIMATION

1. INTRODUCTION

In this chapter we describe briefly the methods adopted to arrive at the solutions of the wave equation in one dimension by well known eikonal approximations [1], successive applications of the Liouville transformations [2], and the elegant matrix formalism developed in the monograph by Fröman and Fröman [3]. We will relate them with the successive approximations, arrived at by methods based on imbedding principles in later chapters.

2. THE EIKONAL EXPANSION

Extending the method of JWKB to solve the Schrödinger Equation [4], we will adopt the optical method of expressing the phase of the solution as a series in terms of some parameters in the equation. In the quantum mechanical case, we use the inverse of λ as the expansion parameter, where $\lambda = 1/\hbar$, which is very large since $\hbar$ Planck's constant is very small. Hence, rewriting the Schrödinger Euqation as (using proper units):*

$$\frac{d^2\Psi}{dx^2} + \lambda^2 k^2 \Psi = 0 \tag{1}$$

k being a function of x, we can write the solution in general in the following manner.

$$\Psi = \exp(\lambda\, S(x)), \text{ with } S(x) = \int_0^x \sum_{n=0}^{\infty} \frac{y_n}{\lambda^n}\, dx' \tag{2}$$

---

* The notation is slightly different from that used in the last chapter.

If we call dS/dx as Y, then it is easy to see that Y satisfies a Riccati equation of the type (Chapter I)

$$\lambda^2 Y^2 + \lambda Y' + \lambda^2 k^2 = 0; \tag{2a}$$

Y can therefore be assumed to be given by a series of the type

$$Y = \sum_{n=0}^{\infty} \frac{y_n}{\lambda^n} \tag{2b}$$

Substituting Equation (2b) in Equation (2a), we arrive at the recursion relations for y_n

$$y_0^2 = -k^2 \text{ that is } y_0 = \pm ik \tag{3}$$

and

$$\sum_{n=0}^{r} y_n y_{r-n} = -\frac{dy_{r-1}}{dx} \quad r = 1, 2, 3, \cdots, \text{etc.} \tag{4}$$

We can solve for successive terms $y_1, y_2, y_3, \cdots$, etc., starting with $y_0 = ik$ or $y_0 = -ik$. Solving for the first four terms, we have

$$y_0 = \pm ik, \qquad y_1 = \frac{y_0'}{2y_0} = \frac{-k'}{2k} \tag{5}$$

$$y_2 = \frac{1}{2}\left(\frac{y_0''}{2y_0^2} - \frac{3y_0'^2}{4y_0^3}\right) = \pm 1/2ik \left(\frac{k''}{2k} - \frac{3k'^2}{4k^2}\right) \tag{6}$$

$$y_3 = -\frac{1}{8}\left(\frac{y_0'''}{y_0^3} - \frac{6y_0'y_0''}{y_0^4} + \frac{6y_0'^3}{y_0^5}\right) = \frac{1}{8}\left(\frac{k'''}{k^3} - \frac{6k'k''}{k^4} + \frac{6k'^3}{k^5}\right). \tag{7}$$

Thus, we have two solutions, corresponding to the two possible values of y_0, one being the complex conjugate of the other. Also, we notice that alternate terms in the phase are real, since we are concerned with real values of k in this discussion. Hence, we have

$$\exp(\lambda S(x)) = \exp\left[\lambda\int\left(y_0 + \frac{y_1}{\lambda} + \frac{y_2}{\lambda^2} + \cdots\right) dx'\right] \tag{8}$$

and if we stop with the first two terms, we have

$$\begin{aligned}\exp(\lambda S(x)) &= \exp\left[\pm i\lambda\int_0^x k\, dx - \int_0^x (k'/2k)\, dx'\right] \\ &= (1/\sqrt{k})\exp\left(\pm\, i\lambda\int_0^x k\, dx'\right)\end{aligned} \tag{9}$$

which is the well-known JWKB solution. If this solution is to satisfy the wave equation exactly, the terms y_2, y_3, etc., must be equal to zero. In view of the recurrence relation (4), we have

$$\begin{aligned} y_1 &= -\,(1/2)(y_0'/y_0) \\ y_2 &= -\,(y_1' + y_1^2)/2y_0 \\ y_3 &= -\,(y_2' + 2y_1y_2)/2y_0, \text{ etc.}\end{aligned} \tag{10}$$

This implies that if y_2 is put equal to zero all the higher terms y_3, y_4, etc., automatically vanish. To achieve this, the profile of the refractive index k(x) should be such that

$$y_1' = -\,y_1^2. \tag{11}$$

Therefore,

$$y_1 = 1/(x + a) = -\,k'/(2k). \tag{12}$$

This means that

$$k = 1/(Cx + D)^2 \tag{13}$$

C and D being constants of integration. Hence, the JWKB solution satisfies the Schrödinger equation exactly for a refractive index profile of the type in Equation (13). Therefore, if we make the transformations of the dependent and independent variables of the Schrödinger equation we can convert the term k(x) approximately to this form, the accuracy of approximations to the actual solution can be successively increased. We will introduce into Equation (1) variables φ and z instead of Ψ and x, defined by the transformation

$$\Psi = z_1^{-1/2}\,\varphi(z) \tag{14}$$

where

$$z_1 = \frac{dz}{dx}, \qquad z_2 = \frac{d^2z}{dx^2}, \text{ etc.} \tag{15}$$

We will determine the independent variable z suitably leading to a good approximation to the solution of Equation (1). It is to be noted that the above transformation preserves the form of the second order differential equation.

Substituting in Equation (1) the transformation (14), we arrive at

$$\varphi'' + (1/z_1^2)[\lambda^2 k^2 - \tfrac{1}{2} \langle z;\, x\rangle]\varphi = 0, \tag{16}$$

where $\langle z;\, x\rangle$ is called the Schwartzian derivative defined by

$$\langle z;\, x\rangle = -2z_1^{1/2}\,\frac{d^2}{dx^2}\left[1/z_1^{1/2}\right] = \left\{\frac{z_3}{z_1} - \frac{3}{2}\,\frac{z_2^2}{z_1^2}\right\}. \tag{17}$$

Let us rewrite Euqation (16) as

$$\varphi'' + \lambda^2 R^2(z)\varphi = 0 \tag{18}$$

and if z can be chosen so that it satisfies the equation so that $R^2(z) = 1$, we have

$$\lambda^2 k^2 = z_1^2\lambda^2 + (1/2)\,\langle z;\, x\rangle. \tag{19}$$

Then Equation (18) has the exact solutions given by

$$\varphi(z) = \exp\,(\pm i\lambda z). \tag{20}$$

An approximate estimate of z can be chosen from the solution of Equation (19), by neglecting the quantity $\langle z;\, x\rangle$, since it is of order $1/\lambda^2$ compared with other terms and since z can be assumed to be slowly varying with respect to x. In

In such a case we can choose $z_1 = k$ and hence

$$z(x) = \int^{x} k \, dx' \tag{21}$$

consequently Ψ can be obtained as

$$\Psi = k^{-1/2} \varphi(z) \tag{22}$$

Equations (21) and (22) are the well-known Liouville transformations also called flattening transformations [2]. This change of variables leads us to the equation

$$\varphi''(z) + [\lambda^2 - (1/2k^2)\langle z; x\rangle]\varphi(z) = 0 \tag{23}$$

$$\varphi''(z) + \left[\lambda^2 - 1/2\left(\frac{k_{xx}}{k^3} - \frac{3}{2}\frac{k_x^2}{k^4}\right)\right]\varphi(z) = 0, \tag{24}$$

where the primes denote differentiating with respect to z. Converting all the differentiations in Equation (24) with respect to the variable z (since k can be found as a function of z), we can rewrite Equation (24) as

$$\varphi'' + \lambda^2\left\{1 - (1/2\lambda^2)\left(\frac{k''}{k} - \frac{k'^2}{2k^2}\right)\right\}\varphi = 0. \tag{25}$$

Equivalently, we can write

$$\varphi'' + \lambda^2\left\{1 + \frac{1}{\lambda^2}\left[\frac{1}{4}\left(\frac{1}{k}\frac{dk}{dz}\right)^2 - \frac{1}{2k}\frac{d^2k}{dz^2}\right]\right\}\varphi = 0 \tag{26}$$

or

$$\varphi'' + \lambda^2\left\{1 + \frac{1}{\lambda^2}\left[\frac{1}{k^{3/2}}\frac{d^2}{dx^2}\left(\frac{1}{k^{1/2}}\right)\right]\right\}\varphi = 0. \tag{27}$$

Neglecting the Schwartzian derivative, the solution Ψ for Equation (1) is

$$\Psi = k^{-1/2} \exp\left(\pm i\lambda \int^{x} k \, dx'\right) \tag{28}$$

which is the JWKB solution. Following the method of Hecht and Mayer [4], an iterative scheme for solving for z in Equation (19) can be obtained. To this end we note an important property of the Schwartzian derivative. If Ψ obeys the equation $\Psi'' + \lambda^2 k^2 \Psi = 0$ and φ obeys $\varphi'' + \lambda^2 R^2 \varphi = 0$, under the transformation $\Psi = (dz/dx)^{-1/2}\varphi$ we can convert φ equation into Ψ equation under the transformation $\varphi = (dx/dz)^{-1/2}\Psi$ if

$$\lambda^2 k^2 = (dx/dz)^{-2}[\lambda^2 R^2 - \tfrac{1}{2} \langle x;\ z\rangle]. \tag{29}$$

Comparing this with Equations (18) and (19) without equating R = 1 we see that

$$\langle z;\ x\rangle = -(dz/dx)^2 \langle x;\ z\rangle \tag{30}$$

To produce an iteration scheme for z let us assume that z is the correct choice with $R^2 = 1$ satisfying the equation

$$\lambda^2 (dz/dx)^2 + (\tfrac{1}{2}) \langle z;\ x\rangle = \lambda^2 k^2(x), \tag{31}$$

while the transformation $\Psi = (d\eta/dx)^{-1/2}\theta(\eta)$ yields the equation

$$\theta'' + \lambda^2 Q^2 \theta = 0, \tag{32}$$

where Q differs from unity only slightly. The transformation from φ equation to θ equation is given by

$$\varphi = (d\eta/dz)^{-1/2}\theta. \tag{33}$$

Hence, we have the following set

$$\lambda^2 (dz/dx)^2 + (1/2) \langle z;\ x\rangle = \lambda^2 k^2 \quad \text{since R is assumed to be equal to unity.} \tag{34}$$

$$\lambda^2 Q^2 (d\eta/dx)^2 + (1/2) \langle \eta;\ x\rangle = \lambda^2 k^2 \tag{35}$$

$$\lambda^2 (dz/d\eta)^2 + (1/2) \langle z;\ \eta\rangle = \lambda^2 Q^2. \tag{36}$$

Hence, from Equations (35) and (36) we have

$$\lambda^2(dz/d\eta)^2 + (1/2)\ \langle z;\ \eta\rangle =$$

$$= (d\eta/dx)^{-2}[\lambda^2 k^2 - (1/2)\ \langle \eta;\ x\rangle]. \tag{37}$$

Since η is very close to the actual solution z we can neglect $\langle z;\ \eta\rangle$ and obtain

$$(dz/d\eta)^2 = (d\eta/dx)^{-2}[k^2 - (1/2\lambda^2)\ \langle \eta;\ x\rangle] \tag{38}$$

If η is labeled the ν-th approximation to z, z in the above equation may be the $(\nu+1)$th approximation, and we have the iterative scheme given by

$$\frac{dz^{\nu+1}}{dz^{\nu}} = \left(\frac{dz^{\nu}}{dx}\right)^{-1} [k^2 - (1/2\lambda^2)\ \langle z^{\nu};\ x\rangle]^{+1/2}. \tag{39}$$

However, we should take care to keep the bracket under the radical sign always positive. This iteration only means that we perform the Liouville transformation successively on the transformed equations obtained at each stage.

The repeated application of this Liouville transformation will yield better and better approximations to the solution of the wave equations in an inhomogeneous medium. This method has been adopted in various fields [7] and it will be described in later chapters.

3. DERIVATION OF THE SOLUTION OF THE SCHRÖDINGER EQUATION USING MATRIX METHODS

Following the exposition by Fröman and Fröman [3], we start with the Schrödinger Equation (1) and make a general transformation (14)

$$z(x) = \lambda \int^{x} q(x')\ dx', \tag{40}$$

where q is a trial function. We then obtain the transformed Equation (16)

$$\varphi''(z) + [1 + \varepsilon(z)]\varphi(z) = 0, \tag{41}$$

where

$$\varepsilon = \frac{\lambda^2 k^2 - \lambda^2 q^2}{\lambda^2 q^2} + \frac{1}{\lambda^2} q^{-3/2} \frac{d^2}{dx^2} (q^{-1/2}) \tag{42}$$

$$= \frac{\lambda^2 k^2 - \lambda^2 q^2}{\lambda^2 q^2} - q^{-1/2} \frac{d^2}{dz^2} (q^{1/2}) \tag{43}$$

$$= \frac{\lambda^2 k^2 - \lambda^2 q^2}{\lambda^2 q^2} + \frac{1}{2\lambda^2}\left\{\frac{1}{q} \frac{d^2}{dx^2}\left(\frac{1}{q}\right) - \frac{1}{2}\left[\frac{d}{dx}\left(\frac{1}{q}\right)\right]^2\right\}. \tag{44}$$

These are the same equations as Equation (23) expressed in different notation.

All the above are equivalent expressions and if ε is small for a suitable choice of q we can neglect it and find the solution φ of equation (41) accurately as

$$\varphi = \exp(\pm iz). \tag{45}$$

If we can express the solution of the second order differential equation with inhomogeneous potential terms, i.e., when ε is not zero, we can write the solution φ in terms of the solutions (45) as

$$\varphi(z) = a_1(z) \exp(iz) + a_2(z) \exp(-iz) \tag{46}$$

and impose the condition

$$\frac{d\varphi}{dz} = ia_1 \exp(iz) - ia_2 \exp(-iz). \tag{47}$$

By this we imply the following conditions on a_1 and a_2

$$\frac{da_1}{dz} \exp(iz) + \frac{da_2}{dz} \exp(-iz) = 0. \tag{48}$$

If we substitute φ into Equation (41), we find that along with Equation (48) we can replace the second order equation by a system of two first order equations given by

$$\frac{da_1}{dz} = (1/2)i\varepsilon(a_1 + a_2 \exp(-2iz)) \tag{49}$$

$$\frac{da_2}{dz} = (-1/2)i\varepsilon(a_2 + a_1 \exp(2iz)). \tag{50}$$

The above set can be expressed as a vector matrix equation

$$\frac{d\vec{a}}{dz} = N(z)\vec{a}. \tag{51}$$

The matrix N(z) can be represented by

$$N(z) = (1/2)i\varepsilon \begin{bmatrix} 1 & \exp(-2iz) \\ -\exp(2iz) & -1 \end{bmatrix} \tag{52}$$

with the following properties for N(z)

$$\mathrm{Tr}\, N(z) = 0 \quad \text{and} \quad \mathrm{Det}\, N(z) = 0. \tag{53}$$

The solution of (51) for suitable initial conditions for $\vec{a}(z)$ can be found as

$$\vec{a}(z) = [F(z, z_0)]\vec{a}(z_0), \tag{54}$$

where

$$[F(z, z_0)] = 1 + \int_{z_0}^{z} N(z_1)\, dz_1 + \int_{z_0}^{z} dz_1 N(z_1) \int_{z_0}^{z_1} dz_2 N(z_2) + \int_{z_0}^{z} dz_1 N(z_1) \int_{z_0}^{z_1} dz_2 N(z_2) \int_{z_0}^{z_2} dz_3 N(z_3) + \cdots. \tag{55}$$

Alternatively, we can write the matrix differential equation

$$\frac{\partial F(z, z_0)}{\partial z} = N(z)F(z, z_0) \tag{56}$$

with initial condition

$$F(z_0, z_0) = 1. \tag{57}$$

Also, we see that

$$\frac{d}{dz} \det F(z, z_0) = \det N(z) \det F(z, z_0) = 0. \tag{58}$$

Since $\det F(z_0, z_0) = 1$, we conclude that $\det F(z, z_0) = 1$ for any z. From Equation (55) we know that

$$F(z_0, z) = [F(z, z_0)]^{-1}. \tag{59}$$

It is a crucial fact that we can factorize the product of matrices $N(z_1)N(z_2) \cdots N(z_n)$ as follows

$$N(z_1)N(z_2)\cdots N(z_n) = (\tfrac{1}{2}i)^n \varepsilon(z_1)\varepsilon(z_2)\cdots\varepsilon(z_n)$$
$$[1 - \exp\{-2i(z_1-z_2)\}] \times [1 - \exp\{-2i(z_2-z_3)\}]\cdots$$
$$[1-\exp\{-2i(z_{n-1}-z_n)\}] \times \tag{60}$$
$$\begin{pmatrix} 1 & \exp(-2iz_n) \\ -\exp(2iz_1) & -\exp\{2i(z_1-z_n)\} \end{pmatrix}.$$

In view of Equation (60), we can easily compute the elements of the matrix and hence

$$F_{11}(z, z_0) = 1 + \int_{z_0}^{z} dz_1 \frac{1}{2} i\varepsilon(z_1) +$$
$$+ \int_{z_0}^{z} dz_1 \tfrac{1}{2}i\varepsilon(z_1)\int_{z_0}^{z_1} dz_2 \tfrac{1}{2}i\varepsilon(z_2) \times$$
$$\times [1-\exp\{-2i(z_1-z_2)\}]+$$
$$+ \int_{z_0}^{z} dz_1 \tfrac{1}{2}i\varepsilon(z_1)\int_{z_0}^{z_1} dz_2 \tfrac{1}{2}i\varepsilon(z_2) \times \tag{61}$$

$$\times [1 - \exp\{-2i(z_1-z_2)\}]\int_{z_0}^{z_2} dz_3 \frac{1}{2}i\varepsilon(z_3) \times$$

$$[1 - \exp\{-2i(z_2-z_3)\}] + \cdots,$$

and

$$F_{22} = F_{11}^* \tag{62}$$

$$F_{12} = \int_{z_0}^{z} dz_1 \frac{1}{2}i\varepsilon(z_1) \exp(-2iz_1) +$$

$$+ \int_{z_0}^{z} dz_1 \frac{1}{2}i\varepsilon(z_1) \int_{z_0}^{z_1} dz_2 \frac{1}{2}i\varepsilon(z_2) \times$$

$$\times [1 - \exp\{-2i(z_1-z_2)\}][\exp\{-2iz_2\}] + \tag{63}$$

$$+ \int_{z_0}^{z} dz_1 \frac{1}{2}i\varepsilon(z_1) \int_{z_0}^{z_1} dz_2 \frac{1}{2}i\varepsilon(z_2) \int_{z_0}^{z_2} dz_3 \frac{1}{2}i\varepsilon(z_3) \times$$

$$\times [1 - \exp\{-2i(z_1-z_2)\}][1 - \exp\{-2i(z_2-z_3)\}]$$

$$\exp\{-2iz_3\} + \cdots,$$

$$F_{21} = F_{12}^* . \tag{64}$$

From the above we obtain the solution φ of Equation (46) and then the solution Ψ of the original Schrödinger equation can be easily seen to be

$$\Psi(x) = a_1(x)f_1(x) + a_2(x)f_2(x) , \tag{65}$$

where

$$f_1(x) = q^{-1/2} \exp\left(+i\lambda \int^{x} q(\xi)\, d(\xi)\right) \tag{66}$$

$$f_2(x) = q^{-1/2} \exp\left(-i\lambda \int^{x} q(\xi)\, d(\xi)\right). \tag{67}$$

Hence, we can write the general solution

$$\Psi = a_1(x)f_1(x) + a_2(x)f_2(x), \tag{68}$$

and a_1 and a_2 can be computed using the values of $F(z, z_0)$ to any degree of accuracy. In the series expansion for the elements of $F(z, z_0)$ matrix, each ε has a dependence $1/\lambda^2$ as seen from Equations (42), (43), (44), and (45). Hence, we can stop with any order of required accuracy in the expressions for a_1 and a_2 and Ψ is a linear combination of two independent solutions

$$\begin{aligned} &= a_1(z_0)[F_{11}(z, z_0) + \exp\{-2iz\} F_{21}(z, z_0)]f_1(z) + \\ &\quad + a_2(z_0)[F_{12}(z, z_0) \exp\{2iz\} + F_{22}(z, z_0)]f_2(z). \end{aligned} \tag{69}$$

The Wronskian of the two parts of the solution on the right hand side of (69) is a constant (equal to -2i) and hence the two parts are linearly independent solutions of the original Schrödinger Equation [3]. In the above, the convergence of the elements of the F matrix has been assumed and various estimates can be obtained. The general term in the series for $F_{11}(z, z_0)$ can be estimated as

$$\begin{aligned} &\left| \int_{z_0}^{z} dz_1\, \tfrac{1}{2} i\varepsilon(z_1) \int_{z_0}^{z_1} dz_2\, \tfrac{1}{2} i\varepsilon(z_2) \Big[1 - \right. \\ &\quad - \exp\{-2i(z_1 - z_2)\}\Big] \int_{z_0}^{z_2} dz_3 \cdots \int_{z_0}^{z_{n-1}} dz_n\, \tfrac{1}{2} i\varepsilon(z_n) \Big[1 - \\ &\quad \left. - \exp\{-2i(z_{n-1} - z_n)\}\Big] \right| \le \frac{1}{2L} \frac{(L\pi)^n}{n!}, \end{aligned} \tag{70}$$

where L is the estimate of

$$\frac{1}{2}|[1 - \exp\{-2i(z_{n-1}-z_n)\}]| \leq L \tag{71}$$

and π is given by

$$\pi = \left| \int_{z_0}^{z} \varepsilon(z_1)\, dz_1 \right| . \tag{72}$$

We will assume $|\exp iz|$ monotonically increases or at least does not decrease as we move from z_0 to z_1 and if k^2 is positive this offers no difficulty at all if we move along the real axis. With these estimates one obtains:

$$|F_{11}(z, z_0) + \exp(-2iz)F_{21}(z, z_0)| \leq \{\exp L\pi\} \tag{73}$$

$$|\exp(2iz)F_{12}(z, z_0) + F_{22}(z, z_0)| \leq \{\exp L\pi\} \tag{74}$$

since path from z_0 to z is along the real axis. Hence, in view of Equation (68) for the Ψ function the two linearly independent parts can be estimated as

$$\begin{aligned}\Psi &= a_1(z_0)f_1(z)\exp(L\pi) + a_2(z_0)f_2(z)\exp(L\pi)\\ &= \Psi_1 + \Psi_2.\end{aligned} \tag{75}$$

The maximum estimate of L can be taken as L = 1 for path along the real axis, for k^2 being greater than 0. Hence, we can write

$$\Psi_1 \sim f_1(1 + \delta), \tag{76}$$

where

$$\delta = [\exp \pi - 1]$$

with

$$\pi = \left| \int_{z_0}^{z} \frac{1}{q^{1/2}} \frac{d^2}{dz^2} \left(\frac{1}{q^{-1/2}}\right) dz \right| . \tag{77}$$

This estimate of the accuracy of the JWKB approximation is that corresponding to the results obtained by Olver [5]. From this standpoint it is customary to study the nature of the solutions near the classical turning points, i.e., near the region where $k^2(x) = 0$, and establish connection formulae for regions below and above the turning points. We will not go into these questions now but turn our attention from the next chapter onwards to methods based on the principles of invariant imbedding to arrive at the solutions of the wave equations in the inhomogeneous region by relating them to the external solutions, where the wave equation has constant k values. We will therefore be concerned with situations in which there are no turning points. The importance of these imbedding techniques lies in the fact that they reduce boundary value problems into initial value problems which facilitate numerical computations to a great extent. Contact with Bremmer series type solutions is easily established, and the method can be generalized to higher order equations and to equations with sources.

4. ASYMPTOTIC BEHAVIOR OF THE SOLUTIONS

If we start with the equation

$$u'' + a^2(t)u = 0 \tag{78}$$

and make the change of variable

$$x = \int_0^t a(t_1)\, dt_1 \tag{79}$$

we have

$$\frac{d^2u}{dx^2} + \frac{a'(t)}{a(t)}\frac{du}{dx} + u = 0. \tag{80}$$

A further change of variable u

$$u = v/a^{1/2} \tag{81}$$

converts the original equation to the form

$$v'' + (1 + b)v = 0. \tag{82}$$

If the initial conditions on v are $v(0) = c_1$, $v'(0) = c_2$ then we are interested in the boundedness of the solutions of (82) as $t \to \infty$.

Let

$$\int_0^\infty |b(t)| < \infty \tag{83}$$

then all the solutions of (82) are bounded as $t \to \infty$ [7]. This can be seen if we write the solution of (82) as

$$v = c_1 \cos x + c_2 \sin x - \int_0^x \sin (x-s) b(s) v(s) \, ds. \tag{84}$$

Taking the absolute values on both sides we have the inequality

$$|v| \leq |c_1| + |c_2| + \int_0^x |b(s)| |v(s)| \, ds. \tag{85}$$

This inequality yields an upper bound for $|v(x)|$ due to a result known as the Gronwell-Bellman inequality [6]. This result can be stated as follows.

If r, c_3, $m \geq 0$ then the inequality

$$r(t) \leq c_3 + \int_0^t m(s) r(s) \, ds \tag{86}$$

implies the inequality

$$r(t) \leq c_3 \exp \left(\int_0^t m(s) \, ds \right). \tag{87}$$

From (86) we see that

$$\frac{r(t)}{c_3 + \int_0^t m(s) r(s) \, ds} \leq 1. \tag{88}$$

Multiplying by m(t) and integrating between 0 and t, we obtain

$$\log\left(c_3 + \int_0^t m(s)r(s)\,ds\right) - \log c_3 \le \int_0^t m(s)\,ds \tag{89}$$

or

$$c_3 + \int_0^t m(s)r(s)\,ds \le c_3 \exp\left[\int_0^t m(s)\,ds\right]. \tag{90}$$

Hence, Equation (87) is proved. Applying this inequality to Equation (85), we have

$$\begin{aligned} |v| &\le (|c_1| + |c_2|) \exp\left[\int_0^x b(r)\,dr\right] \\ &\le (|c_1| + |c_2|) \exp\left[\int_0^\infty b(r)\,dr\right]. \end{aligned} \tag{91}$$

This establishes the boundedness. Returning to Equation (84) we rewrite it as

$$\begin{aligned} v(x) = c_1 \cos x + c_2 \sin x - \Bigg[&\sin x \int_0^x \cos(s)b(s)v(s)ds - \\ &- \cos x \int_0^x \sin(s)b(s)v(s)\,ds\Bigg]. \end{aligned} \tag{92}$$

Since v(s) is bounded and $\int_0^\infty |b(s)|\,ds < \infty$, the integrals

$$a_1 = \int_0^\infty \cos(s)b(s)v(s)\,ds \qquad a_2 = \int_0^\infty \sin(s)b(s)v(s)\,ds \tag{93}$$

are convergent. Hence,

$$\begin{aligned} v(x) &= c_1 \cos x + c_2 \sin x - a_1 \sin x + a_2 \cos x + \\ &\quad + \sin x \int_x^{\infty} \cos(s) b(s) v(s)\, ds - \\ &\quad - \cos x \int_x^{\infty} \sin(s) b(s) v(s)\, ds \\ &= (c_1 + a_2) \cos x + (c_2 - a_1) \sin x + o(1) \end{aligned} \tag{94}$$

as $x \to \infty$. Since the original solution $u = v/a^{1/2}$, we have the asymptotic behavior of the solution of Equation (82) given by

$$\frac{\bar{c}_1 \cos\left(\int_0^t a(t_1)\, dt_1\right) + \bar{c}_2 \sin\left(\int_0^t a(t_1)\, dt_1\right)}{\sqrt{a(t)}} + o\left(\frac{1}{\sqrt{a(t)}}\right), \tag{95}$$

which is in the form of the WKB solutions. Thus making use of the Liouville transformations one can study the asymptotic behavior. The aim is to obtain the term "b" in the transformed equations as small as possible and have suitable properties. If necessary the transformation may be applied over and over to this end.

REFERENCES

[1] Flügge, S., Practical Quantum Mechanics I, Springer-Verlag, Berlin, New York, 1971.
[2] Liouville, J., Journal de Mathematique 2 (1837) 16-35.
[3] Fröman, Nanny and Per Olaf Fröman, JWKB Approximation - Contributions to the Theory, North Holland Publishing Co., New York, 1965.
[4] Hecht, Che and J.E. Mayer, 'Extension of the WKB Equation', Phys. Rev. 106 (1957), 1156.
[5] Olver, F.W.J., 'Error Bounds for the Liouville-Green (or WKB) Approximation', Cambridge Phil. Society Proceedings 57 (1961) 790-810.
[6] Beckenbach, E.F. and R. Bellman, Inequalities, Springer-Verlag, New York, 1961.
[7] Bellman, R., Stability Theory of Differential Equations, Dover Publications, New York, 1969.

CHAPTER III

INVARIANT IMBEDDING

1. INTRODUCTION

In this chapter we introduce the basic ideas of the method of invariant imbedding, relating to the simple case of particle transport in a one-dimensional medium. The non-linear equations for the emergent fluxes are deduced by particle counting techniques as well as by analytical procedures. The importance of conservation laws, positivity of the fluxes, are discussed. A method of linearizing the equations by a method dictated by the physics of the problem is elaborated. In the succeeding chapters, these methods are adapted to the case of wave propagation.

2. INVARIANT IMBEDDING METHOD

There has been within the last few years a large amount of research work, employing the methodology of 'invariant imbedding' [1] - [7] for the analysis of an extensive range of problems, in physics, engineering, astrophysics, nuclear technology and other types of transport phenomena. The basic principle of imbedding was also used earlier [8], [9] under the name of invariance principles in the context of radiation transport. This type of mathematical analysis which leads to functional equations, essentially belongs to the realm of perturbation techniques in which the quantity that is perturbed is the extension or the size of the medium itself instead of any parameter in the equation as in previous chapters. The physical concepts used to arrive at the functional equations provide new insights into the analytic structure of the solutions by taking into account the fundamental processes involved in the model governed by the original differential equations.

Let us now briefly introduce the concept of invariant imbedding and its application to two point boundary value problems. As is well known if we consider a second order equation,

$$u''(z) + k^2(z)u(z) = 0, \quad a \le z \le b \tag{1}$$

with boundary values of u defined at z = a and z = b we may obtain one unique solution or many solutions or no solution at all depending on the boundary values u(a) and u(b) and the extension or length of the medium $\ell = b - a$. In contrast, if we have an initial value problem, we can fairly well make statements regarding the uniqueness and existence of the solutions and more than that the computation of the solutions of initial value problems are devoid of many instabilities that arise in solving the boundary value equations. Moreover, the modern digital computers are admirably well suited to compute the solutions of initial value equations, with great accuracy and speed. Imbedding techniques provide tools to convert boundary value problems into Cauchy problems with initial values. Basically, the method involves generating a 'family' of problems by means of a basic single parameter, while the basic structure of the system remains invariant under the generation of the family. To understand this concept let us take a simple model of particle transport in a one-dimensional rod.

3. THE CLASSICAL APPROACH

Let us consider (a line segment) a rod of length, X, extending from z = 0 to z = X with beams of particles moving in either direction without interaction or collision with each other but interacting with the medium. Let u(z) be the expected flux of the right moving beam and v(z) the average left moving flux where z is a generic point in the medium. Let $\sigma(z)\Delta$ be the probability that there is a collision of a particle with the medium, in traversing a distance Δ, and let us assume that at each collision a fraction f(z) of the particles goes forward and b(z) of the particles is reflected back on the average. With this model, the right moving flux at $(z + \Delta)$ can be related to that at z by the equation

$$u(z+\Delta) = [1-\sigma\Delta]u(z) + \sigma\Delta fu + \sigma\Delta b(z)v(z) + o(\Delta). \tag{2}$$

The above equation in the continuous limit as $\Delta \to 0$ yields the differential equation

$$\frac{du}{dz} = \sigma\{f(z) - 1\}u(z) + \sigma b(z)v(z). \tag{3}$$

A similar analysis gives the equation for v(z) as

$$-\frac{dv}{dz} = \sigma(z)\{f(z) - 1\}v(z) + \sigma(z)b(z)u(z). \qquad (4)$$

This set of equations for the steady state has to be solved under the boundary conditions that unit flux enter the rod at the right end and no particles enter at the left. Moreover, there are no sources or sinks for the particles inside the rod, i.e., we have to solve the set (3) and (4) under the conditions

$$u(0) = 0; \qquad v(x) = 1. \qquad (5)$$

In this approach, we obtain the internal solutions for u and v inside the region $0 \leq z \leq x$.

This type of analysis of the transport of particles in the context of the neutron diffusion problem with various modifications was extensively treated by Ramakrishnan et al [10]. On the other hand, the series of papers by Bellman et al [11] deals with the particle transport concentration of the emergent fluxes instead of the internal solutions. If we are interested only in the emergent flux of particles u(x) and v(0), we can define the functions t(X) and r(X) as:

t(X) = the expected number of particles emerging from the end z = 0, for unit input at the end Z = X; (6)

r(X) = the expected number of particles reflected from the length of the rod, i.e., coming out at Z = X, for unit input at Z = X. (7)

Thus t(X) and r(X) are functions of the length of the rod and we obtain equations governing the changes of emergent fluxes by imbedding this rod of length X in a class of rods of different lengths. This leads to nonlinear equations, with initial conditions.

4. THE INVARIANT IMBEDDING APPROACH FOR PARTICLE TRANSPORT

Keeping the physical parameters of the system to be the same as that in the original model of the rod of length X, we compute the reflected flux emerging from the right end for unit input at that end, for a rod of slightly larger length $(X + \Delta)$, and relate it to the reflected flux from the rod of length X. To this end we adopt the particle counting technique.

If we fix our attention on $r(X + \Delta)$ the expected number of particles per second coming out from the end $Z = X$, (due to unit flux entering per second at this end) we take account of the following physical events.

(a) The input particle at $(X + \Delta)$ in passing through the interval $X \leq Z \leq X + \Delta$ gives rise to on the average $\sigma b\Delta$ particles reflected to the right, contributing to $r(X + \Delta)$. It also produces on the average $f\sigma\Delta$ particles impinging on the subrod of length X.

(b) There may be no collisions in $X \leq Z \leq X + \Delta$ region and $(1 - \sigma\Delta)$ of the input particle along with $f\sigma\Delta$ forward moving particles may act as a source for the subrod of length X. Let us call this source $s = \{\sigma f\Delta + (1 - \sigma\Delta)\}$.

(c) This source of particles will give rise to a flux of $sr(X)$ reflected from the subrod of which a portion may then pass on to the right and emerge at $(X + \Delta)$ thus contributing to the flux $r(X + \Delta)$.

(d) If the out-coming beam at X meets with a collision in the region $X \leq Z \leq X + \Delta$ flux amount of $srb\Delta$ will be incident back into the subrod of length X, and will yield a reflected flux, $sr^2(X)b\Delta(1-\sigma\Delta)$, as the emergent beam at $(X + \Delta)$.

(e) Because of the collision mentioned in (d), a forward flux contribution of $srf\Delta$ will also add up to the emergent flux $r(X + \Delta)$.

If one attempts to calculate further contributions to $r(X + \Delta)$ they will be found to be of order $o(\Delta)$ since they will involve at least two collisions.

Hence, the total contribution to $r(X + \Delta)$ turns out to be

$$r(X+\Delta) = r(X) + \sigma\Delta\{b+2(f-1)r(X) + br^2(X)\} + o(\Delta). \quad (8)$$

Hence, going to the limit $\Delta \to 0$ we have the Riccati differential equations satisfied by $r(X)$:

$$\frac{dr}{dx} = \sigma\{b + 2(f-1)r + br^2\}. \quad (9)$$

A similar analysis leads to the equation for the transmitted average flux $t(X)$ emerging at the end $Z = 0$, due to unit input flux at $Z = X$, and we have

$$\frac{dt(X)}{dX} = \sigma\{(f-1) + br(X)\}t(X). \quad (10)$$

If we imagine the rod to be of zero length, i.e., if $X = 0$, there will be no reflected flux and all of the incident beam will be transmitted. Hence, the initial conditions for Equations (9) and (10) are

$$r(0) = 0 \quad \text{and} \quad t(0) = 1. \tag{11}$$

Here we have assumed the continuity of the parameters. Once the emergent reflection and transmission functions $r(X)$ and $t(X)$ valid for all X in a given interval of interest have been computed numerically or otherwise the internal solutions, $u(Z)$ and $v(Z)$, for boundary values defined in Equation (5) can be written down using the invariance concept. This means

$$u(Z) = r(Z)\, v(Z) \tag{12}$$

$$v(Z) = t(X-Z) + r(X-Z)u(Z). \tag{13}$$

Equations (12) and (13) are a coupled set of algebraic equations which can be solved very easily once we have mapped out $r(X)$ and $t(X)$ as continuous functions of their arguments in the required interval.

The above model can be amplified by assuming that the particles can be found in any one of a finite number of states, $i = 1, 2, \ldots, n$ and, consequently, the cross section for collisions σ_i may be different for each i. Moreover, production of the average number of particles moving forward and those reflected backwards may be represented by elements of matrices (f_{ij}) and (b_{ij}). Also, one may impose the condition that the parameters governing forward production (or scattering) and backward production (or scattering) may be different for the left moving and right moving particles. Hence, one may arrive at general vector matrix type of equations governing the transport of the different species (for details see Reference [2]), and we have a general description of the transport problem,

$$\frac{d\vec{u}}{dZ} = A\vec{u} + B\vec{v} \qquad 0 \le Z \le X \tag{14}$$

$$-\frac{d\vec{v}}{dZ} = C\vec{u} + D\vec{v}, \tag{15}$$

where A, B, C, and D are matrices and the boundary values are

$$\vec{u}(0) = 0 \quad \text{and} \quad \vec{v}(X) = \vec{r} \; ; \quad V_i = \delta_{ik}, \; i=1, 2, \ldots, n \tag{15a}$$

corresponding to an input of a kth species particle at the end Z = X.

Naturally, therefore, we have to think of reflection and transmission functions r_{ij} and t_{ij} corresponding to the input of an ith species and the emergence of jth type of particle, by reflections or transmission.

We consider now the reflection matrix $R_\ell(X)$ for the unit input of left moving flux $\vec{v}$ at the right end Z = X and the transmission matrix $T_\ell(X)$ for unit input at the right end. To tackle this problem, without recourse to the particle counting technique described above, we will consider in place of the set (14) and (15) the following matrix equations,

$$\frac{dU}{dX} = AU + BV \tag{16}$$

$$-\frac{dV}{dX} = CU + DV, \tag{17}$$

where the matrices U and V replace the vectors $\vec{u}(X)$ and $\vec{v}(X)$ in (14) and (15). Since there can be two solutions U_1, U_2, and V_1, V_2 for the above set, we will fix the boundary values for these as

$$U_1(0) = 1 \qquad U_2(0) = 0 \tag{18}$$

$$V_1(0) = 0 \qquad V_2(0) = 1. \tag{19}$$

If the boundary values of the vectors $\vec{u}(0) = d$, $\vec{v}(X) = \vec{r}$, we can express $\vec{u}$ and $\vec{v}$ as

$$\vec{u} = U_1\vec{d} + U_2\vec{b} \tag{20}$$

$$\vec{v} = V_1\vec{d} + V_2\vec{b}. \tag{21}$$

The vector $\vec{b}$ has to be determined by the condition

$$\vec{v}(X) = \vec{r} = V_1(X)\vec{d} + V_2(X)\vec{b} \tag{22}$$

if we also have the condition

$$\vec{d} = 0, \quad \text{then} \quad \vec{b} = \left[V_2(X)\right]^{-1} \vec{r}. \tag{23}$$

Hence, the reflected flux at Z = X is given by

$$\vec{u}(X) = U_2(X)[V_2(X)]^{-1}\vec{r} = R_\ell \vec{r}. \tag{24}$$

Differentiating R_ℓ with respect to X and making use of Equations (16) and (17), we have the matrix Riccati equation

$$R_\ell' = B + AR_\ell + R_\ell D + R_\ell C R_\ell \tag{25}$$

with the initial condition

$$R_\ell(0) = 0; \tag{26}$$

similarly the transmission matrix $T_\ell(X)$ is determined by

$$\vec{v}(0) = [V_2(X)]^{-1}\vec{r} = T_\ell(X)\vec{r}. \tag{27}$$

The differential equation satisfied by the transmission matrix $T_\ell(X)$ making use of the set (16) and (17) is given by

$$\frac{dT_\ell}{dX} = T_\ell(CR_\ell + D) \tag{28}$$

with the initial condition

$$T_\ell(0) = 1. \tag{29}$$

This is for the input of particles from the right-hand side and variation of the length of the medium at the right end. Similar equations for the emergent flux can be obtained for input at the left end. The equations obtained by varying the length of the medium at the end opposite to the input end form an entirely different set of equations [2].

5. RICCATI TRANSFORMATIONS

Now we can eliminate the particle transport idea used for framing the set of Equations (3), (4) with given boundary values and start with a set of coupled linear differential equations and study them analytically by applying the Riccati transformation [4].

Let us consider the scalar set of equations,

$$\frac{\partial u(Z, X)}{\partial Z} = a(Z)u(Z, X) + b(Z)v(Z, X) + e(Z) \tag{30}$$

$$-\frac{\partial v(Z, X)}{\partial Z} = c(Z)u(Z, X) + d(Z)v(Z, X) + f(Z), \tag{31}$$

subject to the boundary conditions,

$$u(0, X) = \alpha; \qquad v(X, X) = \beta. \tag{32}$$

u and v are functions of the point Z, as well as the length of the medium X. We assume that the coefficients a, b, c, d, e, and f, the stated boundary conditions (32), and the interval length X, are such that for all $X \leq \Lambda$ the system has a unique solution. We can easily see that the set (30) and (31) correspond to a second order linear differential equation with forcing terms [4].

We introduce the generalized Riccati transformation

$$u(Z, X) = R_\ell(Z)v(Z, X) + T_r(Z)u(0, X) + S_1(Z) \tag{33}$$

and the recovery transformation

$$v(0, X) = T_\ell(Z)v(Z, X) + R_r(Z)u(0, X) + S_2(Z). \tag{34}$$

These transformations along with the original set of equations lead to the following set of first order equations with initial conditions,

$$R_\ell'(Z) = b(Z) + [a(Z) + d(Z)]R_\ell + c(Z)R_\ell^2 \tag{35}$$

$$T_r'(Z) = [a(Z) + cR_\ell(Z)]T_\ell \tag{36}$$

$$S_1'(Z) = [a(Z) + cR_\ell(Z)]S_1 + f(Z)R_\ell + e(Z) \tag{37}$$

with the set of initial conditions

$$R\ (0) = 0; \quad T_r(0) = 1 \ \text{ and } \ S_1(0) = 0. \tag{38}$$

Simultaneously, we can form another set, using Equations (33) and (34),

$$T_\ell'(Z) = [d(Z) + c(Z)R_\ell(Z)]\,T_\ell(Z) \tag{39}$$

$$R_r'(Z) = c(Z)T_\ell(Z)T_r(Z) \tag{40}$$

$$S_2'(Z) = [c(Z)S_1 + f(Z)]\,T_\ell(Z) \tag{41}$$

with initial conditions

$$T_\ell(0) = 1; \quad R_r(0) = 0; \quad S_2(0) = 0. \tag{42}$$

The above set of six equations is independent of boundary values α and β and allows us to study the solution when α, β and the medium length X are changed. The physical significance of R_ℓ, R_r, T_ℓ, T_r and the source-produced fluxes S_1 and S_2 are easily understood. In terms of these functions the internal solutions u(Z) and v(Z) are given as

$$v(Z, X) = T_\ell^{-1}(Z)\Big\{T_\ell(X)\beta + \alpha[R_r(X) - R_r(Z)] + \\ + [S_2(X) - S_2(Z)]\Big\} \tag{43}$$

$$u(Z, X) = R_\ell(Z)v(Z, X) + \alpha T_r(Z) + S_1(Z). \tag{44}$$

Equation (43) is obtained from Equation (34) after substituting boundary values.

6. LINEARIZATION AND SOLUTION OF THE RICCATI EQUATIONS

In the above sets of equations both in the matrix and scalar case, (25) and (28), (35) through (42), the functions describing the emerging fluxes always satisfy nonlinear differential equations. Various techniques like quasi-linearization, etc., [12] can be adopted to solve them. However, for the Riccati type equations, connected with transport theory, we can achieve a linearization, dictated by the physics of the problem itself. Different characterizations of the underlying process may lead to different types of linearization. Once we arrive at the linearization we may discard the physical picture, and calculate the solutions of the original problem in an iterative fashion. This has been widely used in the field of radiation dosimetry [13], [14]. If we consider the transport problem of a set of particles described by the vector-matrix equation for a one-dimensional medium, we have the set of linear Equations (14) and (15). We can easily see that linearization is built into the system automatically if we concentrate on the reflected and transmitted flux of particles which have had a specified number of reflections or backsatterings inside the medium. Each element $r_{ij}(X)$ of the reflection matrix, R, is the reflected flux of jth type due to a unit input of ith type at Z = X. Assuming the boundary $\vec{u}(0) = 0$ and $\vec{v}(X) = \vec{r}$ we can label each reflected beam as $r_{ij}^{2n+1}(X)$ with n = 0, 1, 2, ... where super fix indicates that the beam has had 2n+1 backscatterings inside the medium and has come out in the jth state. It is easy to see that for emergence as the reflected beam, at the input end, the number of backscatterings has to be odd. Similarly, we can characterize the elements t_{ij} of the transmission matrix as $t_{ij}^{2n}(X)$ with n = 0, 1, 2, ... denoting that an even number of backscatterings have been suffered by the beam to come out at the end Z = 0, for the unit input at Z = X. If we write down the contributions to the elements $r_{ij}^{2n+1}(X+\Delta)$, which represents the flux, emerging from the rod of length (X+Δ) in the state j produced by the input flux of the input beam in state i after undergoing (2n+1) backscatterings inside the medium, and relate it to the quantity $r_{ij}^{2n+1}(X)$, emerging from a rod of length X, we have by the particle counting technique

$$\begin{aligned} r_{ij}^{2n+1}(X+\Delta) = {} & b_{ij}\delta_{n0}\Delta + \sum_{\substack{k \\ i\neq k}} a_{ik} r_{kj}^{2n+1}(X)\Delta + \\ & + \sum_{\substack{\ell \\ \ell\neq j}} r_{i\ell}^{2n+1}(X) d_{\ell j}\Delta + (1-a_{ii}\Delta) r_{ij}^{2n+1}(X)(1-d_{ij}\Delta) + \\ & + \sum_{m=0}^{n-1} \sum_{\ell,k} r_{i\ell}^{2(n-m)-1}(X) c_{\ell k}\Delta r_{kj}^{2m+1}(X), \end{aligned} \tag{45}$$

where $b_{ij}\Delta$ is the probability that the unit v flux of ith type is reflected back as a jth type particle in traversing the distance $(X+\Delta)$ to X of the medium $c_{ij}(\Delta)$ is the corresponding quantity for the u flux . $a_{\ell k}\Delta$ and $d_{\ell k}\Delta$ are the probabilities for forward scattering of u and v fluxes, the incident ℓ-th type particle being converted into kth type, in traversing the medium between X and $X + \Delta$. $(1-a_{jj}\Delta)$ and $(1-d_{jj}\Delta)$ are the probabilities of the right going u and left going v fluxes to proceed in their original directions without any transformation in the distance Δ of the medium. We assume that a_{jj} and d_{jj} as positive quantities and we go to the limit as Δ goes to zero. Hence, the (2n+1)th backscattered reflected beam emerging from the rod satisfies the matrix equation

$$R'_{2n+1} = B\delta_{n0} + AR_{2n+1} + R_{2n+1}D + \sum_{m=0}^{n-1} R_{2(n-m)-1}C\,R_{2m+1}. \qquad (46)$$

If we sum over n from 0 to ∞ on both sides we revocer the original equation for R (Equation (25), since it is the sum of all orders of reflected beams that constitute the total beam.

$$R = \sum_{n=0}^{\infty} R_{2n+1} \qquad (47)$$

The matrix A and D have as their elements diagonal elements $-a_{jj}$ and $-d_{jj}$ which are negative. We can assume A and D as functions of Z, i.e., the point at which the scatterings take place or we can take them to be constants. We have considered a situation in which the total number of particles is conserved. The medium is lossless. If there is dissipation, we can include it too by introducing a sink in the medium and equations can be written down for the average dissipation per unit time [13]. The initial condition for R_{2n+1} of any order n is given by

$$R_{2n+1}(0) = 0. \qquad (48)$$

Similarly, we can write equations for T_{2n} of any order n, from Equation (28) as

$$T'_{2n} = T_{2n}D + \sum_{m=0}^{n-1} T_{2(n-m)-2}C\, R_{2m+1} \qquad (49)$$

and the initial condition is given by

$$T_{2n} = \delta_{n0}, \qquad (50)$$

where δ_{n0} is the Kronecker delta function.

Summing n on both sides of Equation (49), we revocer the original Equation (28) for T the total transmitted flux matrix since we have

$$T = \sum_{n=0}^{\infty} T_{2n}. \qquad (51)$$

In the above, R_{2n+1} and T_{2n} refer to left moving beams. We have suppressed the index ℓ, for these functions. It is to be noted that Equation (47) and Equation (49) can be solved successively, starting with the value n = 0. In Equation (46) for n = 0 we do not have the last term, which is bilinear in R_{2n+1}, and when we go to higher values of n we find the last term contains products of terms, which have been already computed at earlier stages. Thus, the successive equations are effectively linear and can be analytically solved for small values of n. Similar linearization is also effected in the case of the equation for the transmission matrix of finite orders T_{2n}.

The nature of the functions R_{2n+1} and T_{2n} can be seen more explicitly if instead of the differential equations we describe (46) and (49) by integral equations. For the total reflected matrix R, the integral equation representation is given by (assuming A and D to be constants)

$$R(X) = \int_0^X e^{A(X-t_1)}\Big[B(t_1) + R(t_1)C(t_1)R(t_1)\Big]e^{D(X-t_1)}\, dt_1 \qquad (52)$$

The above can be interpreted as follows. The unit input at Z = X moving as a left going beam gets propagated forward up to t_1 by the operator $e^{D(X-t_1)}$ and either gets reflected at t_1 by $B(t_1)$ and gets propagated up to X as a right going beam by $e^{A(X-t_1)}$ or comes back reflected from the rod of length t_1 as a right going beam with the value $R(t_1)e^{D(X-t_1)}$ gets reflected back by the operation of $c(t_1)$ and again comes back as $R(t_1)C(t_1)R(t_1)e^{D(X-t_1)}$. This gets propagated to $Z = X$ by the operation of $e^{A(X-t_1)}$. Integration over t_1 allows the point t_1 to lie anywhere from 0 to X. With this interpretation of the different terms in Equation (52) it is easy to write down the representation for $R_{2n+1}(X)$ as

$$R_{2n+1}(X) = \int_0^X e^{A(X-t_1)}[B\delta_{n0} + \sum_{m=0}^{n-1} R_{2(n-m)-1}(t_1)C(t_1)R_{2m+1}(t_1)] \times \times e^{D(X-t_1)}dt_1. \tag{53}$$

Similarly, for T_{2n} we have the representation

$$T_{2n}(X) = \delta_{n0}e^{DX} + \sum_{m=0}^{n-1} \int_0^X T_{2(n-m-1)}(t_1)C(t_1)R_{2m+1}(t_1)e^{D(X-t_1)} dt_1. \tag{54}$$

The interpretation of the above equations is quite clear. If we differentiate Equations (53) and (54) with respect to X, the length of the medium and sum over all values of n, we recover the corresponding differential Equations (25) and (28) obtained earlier.

To be associated with a transport process, coefficient matrices A and D (which we have assumed for the above expressions (52), (53) and (54) to be constants, matrices independent of t) must have specific structure. We can prove the positivity of R and T from the order of scattering analysis we have carried out, by imposing certain conditions on A, B, C, and D. Even if the diagonal elements of A and D are negative, e^A and e^D are positive if all the nondiagonal elements are positive. From the integral representation (53), R_1 is positive definite if B has positive definite elements, (provided $X > 0$) and the integral is bounded for finite X.

Hence, substituting R_1 to evaluate R_3 from (53) we find that R_3 can be seen to be positive definite if C has positive definite elements since R_1 is positive definite. Hence, each order of R can be proved to be positive and hence the total R. A similar proof will hold in the case of T.

Having obtained the emergent functions relating to beams that have undergone a specified number of backscatterings inside the medium, it is possible to express the $\vec{u}$ and $\vec{v}$ solutions at any given point inside the medium, labelling them by the number of backscatterings they have undergone until they arrive at point Z. Taking into account the boundary values of $\vec{u}$ and $\vec{v}$ (15a)

$$\vec{u}_{2n+1}(Z) = \sum_{r=0}^{n} R_{2r+1}(Z)\vec{v}_{2(n-r)}(Z) \tag{55}$$

and

$$\vec{v}_{2n}(Z) = T_{2n}(X\_Z)\vec{r} + \sum_{r=0}^{n-1} R_{2r+1}(X\_Z)\vec{u}_{2(n-r)-1}(Z). \tag{56}$$

Naturally the total solutions $\vec{u}$ and $\vec{v}$ are

$$\vec{u}(Z) = \sum_{n=0}^{\infty} \vec{u}_{2n+1}(Z), \quad 0 \le Z \le X \tag{57}$$

$$\vec{v}(Z) = \sum_{n=0}^{\infty} \vec{v}_{2n}(Z), \quad 0 \le Z \le X \tag{58}$$

Thus computing R and T up to a particular order n will enable us to find the internal solutions, accurate up to that order. And if the nature of the problem is such that the values of R and T successively decrease with increasing orders of n we can obtain the internal solutions up to any desired order of accuracy.

7. CONSERVATION RELATIONS

In this section we consider the equations for R and T matrices and establish the boundedness and positivity of the solutions basing ourselves on the principle of conservation of the particles implied by the nature of the problem. The particles get scattered in the medium and there is no loss since there are no other sinks in the medium. They only get converted into other types, moved forward or turned back. Hence, the total input into the rod of length X should equal the total number emerging from the rod from both sides. The equations for R and T are given by (25) and (28)

$$R'_\ell = B + AR_\ell + R_\ell D + R_\ell CR_\ell \qquad R_\ell(0) = 0 \tag{59}$$

and

$$T'_\ell = T_\ell(CR_\ell + D) \qquad T_\ell(0) = 1 \tag{60}$$

R_ℓ and T_ℓ are the reflection and transmission matrices for the case when the input is unit flux at the right end Z = X. Let us redefine the elements of these matrices involved as follows. For the v beam $b_{ij}\Delta$ is the backscattering probability meaning thereby that a jth particle traversing Δ of the medium turns back as the ith particle. $d_{ij}\Delta$ is the forward scattering probability when traversing Δ, when $i \neq j$.
However, when $i = j$ the probability for forward motion is $1 - d_{jj}\Delta$. Similar definitions apply to the elements of the matrices A and C which refer to the u beam. B and C are non-negative matrices while A and D contain negative elements along the diagonal. Local conservation concerning each beam implies the relations

$$M(A + C) = 0; \qquad M(B + D) = 0, \tag{61}$$

where M is the matrix

$$M = \begin{bmatrix} 1 & 1 & 1 & 1 & \cdots & 1 \\ 0 & 0 & 0 & \cdot & \cdots & 0 \\ \cdot & \cdot & \cdot & \cdot & \cdots & \cdot \\ 0 & 0 & 0 & \cdot & \cdots & 0 \end{bmatrix}. \tag{62}$$

We now expect from the physical nature of the problem that the emergent fluxes will satisfy the conservation law

$$M(R_{\ell} + T_{\ell}) = M \tag{63}$$

which is a translation of the fact that the total input is equal to the total output for $X \geq 0$. We will forget the suffix ℓ and differentiate the quantity $M(R + T)$ and obtain the equation for $[M(R + T)]'$

$$\{M(R + T)\}' = M(B + AR + RD + RCR) + MT(CR + D). \tag{64}$$

This equation after some manipulations can be written as

$$\{M(R + T)\}' = M(B + D) + M(R + T - 1)D + M(R + T - 1)CR + M(A + C)R. \tag{65}$$

Assuming R to be known if we call the quantity $Z = M(R + T - 1)$ we find it satisfies the equation (in view of the local conservation equations)

$$Z' = ZD + ZCR$$
$$Z(0) = 0. \tag{66}$$

One solution of Equation (66) is $Z = 0$ and uniqueness shows that it is the only solution within the interval for which R and T exist. Hence, the relation $M(R + T) = M$ holds. Therefore, R and T are bounded. The elements of R and T are shown to be non-negative from the integral representation of the solution for R and T in Equations (52), (53) and (54). In fact, we arrived earlier at the non-negativity from the representation of the finite order solutions. Thus the solutions of (59) and (60) exist and are unique for finite $X \geq 0$. If there are sinks in the medium then absorption probabilities can be introduced and the quantity that is conserved is the total loss in the medium plus the amount of emergent fluxes. The analysis of all these problems can be found in detail in Reference [15].

Thus in this chapter, we have studied the basic concept of the imbedding principle as applied to the problem of particle transport inside a medium. We will continue to apply this approach to the study of wave propagation and obtain the Bremmer series type of solutions which exactly correspond to the finite order of solutions we have discussed above.

8. SCATTERING MATRIX FORMALISM

As a digression here we want to refer to an interesting aspect of transmission line theory of two port networks. Analogies very similar to propagation of a disturbance through an obstacle can be imagined. The obstacle can be described by a scattering matrix which specifies the transmission and reflection coefficients, for the left moving and right moving disturbances. T_ℓ and R_ℓ are the complex amplitudes of a left moving beam emerging from the medium for unit input at the right end $z = X$ of a rod extending from $z = 0$ to $z = X$. Similarly, one can define T_r and R_r for the right moving disturbance. The matrix S given below can be assumed to represent the medium with regard to its reflection and transmission properties.

$$S = \begin{bmatrix} T_r & R_\ell \\ R_r & T_\ell \end{bmatrix} . \tag{67}$$

This is called scattering matrix by Redheffer [16] who arrived at the functional equations in the context of transmission line theory. In the case of lossless networks since in the case of wave propagation the power is proportional to $(\text{amplitude})^2$ we have

$$|T_\ell|^2 + |R_\ell|^2 = 1 \quad \text{and} \quad |T_r|^2 + |R_r|^2 = 1. \tag{68}$$

For dissipative networks the equality signs in the above have to be replaced by $\leq$. To visualize the function of the scattering matrix in network theory we can imagine a four terminal linear network as shown in diagram (1) with inputs z_1, and z_4 and outputs z_3 and z_2. Then these amplitudes are related as

$$\begin{pmatrix} z_3 \\ z_2 \end{pmatrix} = S\begin{pmatrix} z_1 \\ z_4 \end{pmatrix} \tag{69}$$

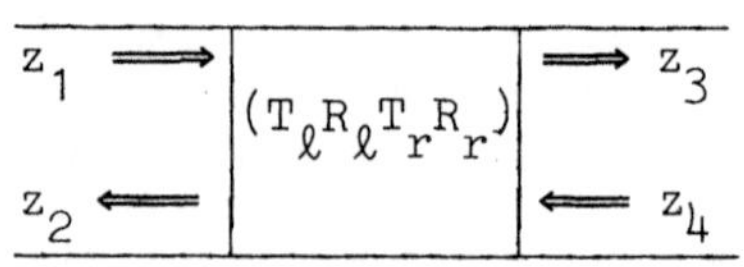

Fig. 1.

where S is the scattering matrix of Equation (67) defining the network. If we now juxtapose two networks S and S_1 and call the amplitudes z_5 and z_6 coming out and going into the right end of the second obstacle S_1 (Figure 2)

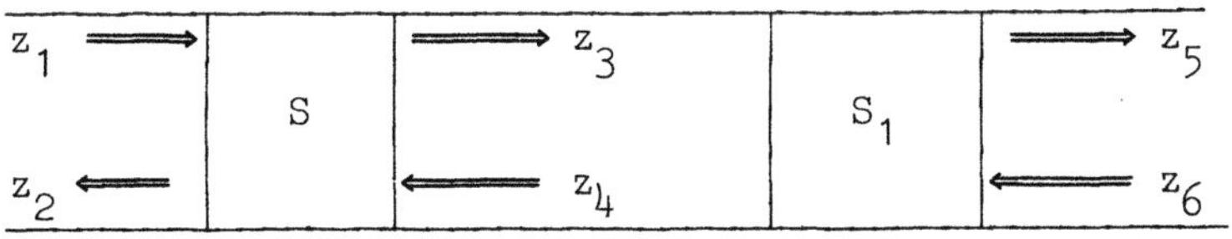

Fig. 2.

we can relate these to the amplitudes of the ingoing and outgoing amplitudes z_4 and z_3 for the right end of the first obstacle S (see diagram). Hence,

$$\begin{pmatrix} z_5 \\ z_4 \end{pmatrix} = \begin{pmatrix} T_{r,1} & R_{\ell,1} \\ R_{r,1} & T_{\ell,1} \end{pmatrix} \begin{pmatrix} z_3 \\ z_6 \end{pmatrix} = S_1 \begin{pmatrix} z_3 \\ z_6 \end{pmatrix}. \tag{70}$$

Where the matrix S_1 specified by the set of complex numbers $(T_{\ell,1}, R_{\ell,1}, T_{r,1}, R_{r,1})$ defines the second network. Utilizing equations (67), (69) and (70) we can see that

$$\begin{pmatrix} z_5 \\ z_2 \end{pmatrix} = \mathcal{S} \begin{bmatrix} z_1 \\ z_6 \end{bmatrix} \tag{71}$$

$$\mathcal{S} = \begin{pmatrix} T_{r1} A T_r & R_{\ell,1} + T_{r,1} R_\ell A T_{\ell,1} \\ R_r + T_\ell R_{r,1} A T_r & T_\ell A T_{\ell,1} \end{pmatrix} \tag{72}$$

with

$$A = (1 - R_\ell R_{r,1})^{-1}. \tag{73}$$

The scattering matrix for the composite obstacle is given by the star product operation defined by Redheffer [16], i.e.:

$$\mathcal{S} = S * S_1. \tag{74}$$

This star product method as defined by Equations (70), (71), (72), and (73) for obtaining the composite matrix takes into account all possible multiple scatterings between the two constituent obstacles in giving rise to the emerging beams at both ends of the composite obstacle or network. We note that this star product is quite different from that of a matrix product.

9. Homogeneous Anisotropic Media Forming an Obstacle

Let us consider an obstacle medium with different reflection and transmission properties for the beams traveling in opposite directions. Since such a medium of thickness (x + y) can be thought of as a medium of thickness x adjacent to a medium of thickness y transmission and reflection coefficients can be computed for unit input amplitude, at the left end of the medium x, using the star product method. Since the transmission and reflection coefficients are functionals of the extent of each medium, let us specify the medium x by the scattering matrix

$$S = \begin{pmatrix} T_r(x) & R_\ell(x) \\ R_r(x) & T_\ell(x) \end{pmatrix} \tag{75}$$

and the medium y by

$$S_1 = \begin{pmatrix} T_r(y) & R_\ell(y) \\ R_r(y) & T_\ell(y) \end{pmatrix}. \tag{76}$$

Hence, using the concept of the star product (71), (72), and (73) we easily see for the total medium (x + y)

$$R_r(x + y) = R_r(x) + \frac{T_r(x)T_\ell(x)R_r(y)}{1 - R_\ell(x)R_r(y)} \tag{77}$$

$$T_r(x + y) = \frac{T_r(x)T_r(y)}{1 - R_\ell(x)R_r(y)}. \tag{78}$$

Similar equations can be written down for $R_\ell(x + y)$ and $T_\ell(x + y)$ for incidence at the right end. We automatically have the initial conditions

$$T_r(0) = T_\ell(0) = 1; \quad R_r(0) = R_\ell(0) = 0. \tag{79}$$

Let us assume that

$$R_r'(0) = c; \; R_\ell'(0) = b; \; T_r'(0) = a; \; T_\ell'(0) = d. \tag{80}$$

It is easily seen that $R_r'(0) = c$ and $R_\ell'(0) = b$ corresponding to the fact that $c\Delta$ and $b\Delta$ are the reflected fractions of the beam moving to the right and moving to the left, respectively. From Equation (77) we have

$$\frac{R_r(x + y) - R_r(x)}{y} = \frac{T_r(x)T_\ell(x)[R_r(y)/y]}{1 - R_\ell(x)R_r(y)} \tag{81}$$

Assuming y to be small and going to the limit $y \to 0$, we have, using Equations (79), (80) and (81).

$$\frac{dR_r}{dx} = cT_r(x)T_\ell(x). \tag{82}$$

This is the case of imbedding in which the variation of the medium is done at the end farther from the incident end. Similarly, the equations for the other quantities can be written down as

$$\frac{dT_r}{dx} = T_r(cR_\ell + a) \tag{83}$$

$$\frac{dT_\ell}{dx} = T_\ell(cR_\ell + d) \tag{84}$$

and

$$\frac{dR_\ell(x)}{dx} = b + (a + d)R_\ell + cR_\ell^2. \tag{85}$$

These are the same equations as those obtained by the imbedding techniques described in earlier sections. This concept of the star product has proved very useful in devising very ingenious methods such as the doubling techniques, etc., in radiative transfer theory [17].

REFERENCES

[1] Bellman, R., R. Kalaba, and G.M. Wing, 'Invariant Imbedding and Mathematical Physics - I: Particle Processes', Journal of Mathematical Physics 1 (1960), 280-308.

[2] Bellman, R. and G.M. Wing, Introduction to Invariant Imbedding, John Wiley & Sons, Inc., New York, 1974.

[3] Meyer, G.H., Initial Value Methods for Boundary Value Problems, (Theory and Applications of Invariant Imbedding), Academic Press Inc., New York, 1972.

[4] Scott, M.R., Invariant Imbedding and Its Applications to Ordinary Differential Equations, Addison-Wesley Publishing Co., Inc., 1973.

[5] Kalaba, R. and J. Casti, Imbedding Methods in Applied Mathematics, Addison-Wesley Publishing Co,m Inc., 1973.

[6] Bellman, R. and R. Kalaba, 'Transport Theory and Invariant Imbedding', American Math. Soc. Symposium, Providence, R.I. 1961.

[7] Shimuzu, A. and K. Aoki, Application of Invariant Imbedding to Reactor Physics, Academic Press, Inc., New York, 1972.

[8] Chandrasekar, S., Radiative Transfer, Oxford, 1950.

[9] Ambarzumian, V.A., 'Diffuse Reflection of Light by a Foggy Medium', Compt. Rendus Academic Science, U.S.S.R., 38, 1943, p. 229.

[10] Ramakrishnan, A.R. Vasudevan, and P. Rajagopal, 'Ambigenous Stochastic Processes', Journal of Mathematical Analysis and Applications 1 (1960), 145-162.

[11] Bellman, R., R. Kalaba, and G.M. Wing:
(a) 'On the Principle of Invariant Imbedding and Neutron Transport Theory - I', Journal of Mathematics and Mechanics 7 (1958), 149.
(b) 'On the Principle of Invariant Imbedding and Neutron Transport Theory - II', Journal of Mathematics and Mechanics 7 (1958), 741.

[12] Bellman, R. and R. Kalaba, Quasilinearization and Nonlinear Boundary Value Problems, American Elsevier Publishing Co., Inc., New York, 1965.

[13] Bellman, R., R. Vasudevan, and S. Ueno, 'On the Matrix Riccati Equations of Transport Processes', Journal of Mathematical Analysis and Applications 44 (1973), 472-481.

[14] Bellman, R., R. Vasudevan, and S. Ueno, 'Invariant Imbedding and Radiation Dosimetry - I: Finite Order

Scattering and Transmission Function', Mathematical Biosciences 14 (1972), 235-254.
[15] Bellman, R., K.L. Cooke, R. Kalaba, and G.M. Wing, 'Existence and Uniqueness Theorems in Invariant Imbedding - I: Conservation Principles', Journal of Mathematical Analysis and Applications 10 (1965), 234-244.
[16] Redheffer, Raymond, Novel Uses of Functional Equations.
(a) Journal Rational Mech. and Analysis 3 (1954), 271.
(b) Article 12 in Modern Mathematics for the Engineers, E.F. Beckenbach (ed.), McGraw-Hill Book Company, New York, 1961.
[17] Vasudevan, R. and R. Bellman, 'Reflection Functions from a Double Layer - A New Approach', Astrophisics and Space Science 28 (1974), 115.

CHAPTER IV

APPLICATION TO THE WAVE EQUATION

1. INTRODUCTION

In this chapter we will take up the study of the wave equations in one dimension and study the propagation of the wave in a region with inhomogeneous properties of refractive index by analyzing the reflection and transmission functions for the region. In the previous chapter we studied these functions in the context of particle transport. Similar studies carry over in the case of wave propagation. An order of scattering analysis of the emergent and internal solutions leads to Bremmer series [1] solutions under certain conditions.

The functional equations based on the imbedding principle in this case may be arrived at either by a 'wavelet-counting' method (similar to particle counting techniques) or by analytical means. For the counting technique we analyze the events that occur to the incident wave within a small region Δ at one end of the medium taking into account the reflection and transmission coefficients for the amplitude of the wave which essentially depends on the relative refractive index of the region Δ of the medium. We also add the phases to the wave when it traverses the stratum Δ, the phases being determined by the wave number within that region. Thus, we adopt the principle of localization which was the basic tenet, in treating a pile of plates or a multilayer of thin optical films elaborated in Chapter I.

This principle [2] may be stated thus: A plane wave traveling through an inhomogeneous medium proceeds as if there were an instantaneous reflection and transmission at each interface of a stratum, between z and z + Δ. These reflections and transmissions occur as if the stratum (z, z + Δ) were actually seme-infinite with a homogeneous refractive index k(z) for any point inside the stratum. This means that the behavior of the wave in any small segment depends entirely on the physical properties of that segment. This localization principle has been tacitly assumed in the particle processes described earlier. Thus, one may trace a single wave accounting for its reflections and transmissions at each interface ahead of it as if the material behind that interface is semi-infinite with a homogeneous wave number pertaining to the small region ahead [3].

Let us now consider a medium extending in half space $-\infty < z < x$ characterized by the wave number $k_1 > 0$. Let the half space $x < z < \infty$ have a wave number $k_2 > 0$. We have an incident wave traveling to the left from $z = \infty$ impinging on the interface $z = x$. Part of the wave might be reflected back and part transmitted to the left at the separation. We are considering a steady beam.

The equations describing the wave function in two regions are

$$\frac{d^2\Psi}{dz^2} + k_1^2\Psi = 0; \quad -\infty < z < x \tag{1}$$

$$k_1 > 0, \quad k_2 > 0 \quad \text{(constants)}$$

$$\frac{d^2\Psi}{dz^2} + k_2^2\Psi = 0; \quad x < z < \infty. \tag{2}$$

The solutions in the two regions are

$$\Psi(z) = e^{-ik_2(z-x)} + \xi(x)e^{ik_2(z-x)} \quad ; \quad x < z < \infty \tag{3}$$

$$\Psi(z) = \tau(x)e^{ik_1(z-x)} \quad ; \quad -\infty < z < x \tag{4}$$

where $\xi(x)$ and $\tau(x)$ are the reflection and transmission factors for unit amplitude of the incident wave at the separation.

The joining conditions at the interface require that Ψ and Ψ' at $z = x$ are continuous and this leads to the evaluation of ξ and τ at an interface (see Chapter I)

$$\xi = \frac{k_2 - k_1}{k_1 + k_2} \tag{5}$$

$$\tau = \frac{2k_2}{k_1 + k_2}. \tag{6}$$

If instead of the two media treated above we have a multimedia stacked as a pile of plates one after another with wave numbers k_i, i=1, 2, ..., N + 1 different from each other, we can by wavelet counting, sum up the contributions due to multiple reflections and transmissions, arrive at the total reflection ξ_{N+1} from the interface at $z = x_{N+1}$ basing ourselves on the principle of localization. In computing the wave incident at the interfaces, the phases due to crossing the stratum $(x_{N+1} - x_N)$ either way is to be taken into account and ξ_{N+1}, the reflection factor due to (N+1) layers can be found to be related to ξ_N the reflections due to the N layers behind. Thus the actual value of ξ_N can be computed by iteration, starting with the reflection factor at the first interface. Similarly the total transmission τ_{N+1} through (N+1) layers can be obtained, if we know ξ_N[4]. In Chapter I we dealt with these types of procedures. Naturally, we want to consider the reflection and transmission from a finite medium with continuously varying wave numbers. This medium is juxtaposed between media of constant wave numbers say k_1 and k_2. The obvious method is to consider the inhomogeneous medium as a multilayer of homogeneous strata, and go to the limit with proper criteria for the pointwise variation of the wave numbers.

2. A CONTINUOUS MEDIUM PROBLEM

We start with the wave equation in a region $x_0 < z < x$ characterized by a wave number which is positive and varying continuously with z. Let this region extend from x_0 to $-\infty$ with a constant wave number k_0 and from x to ∞ with another constant value k_1 for the wave number. In particular we demand the following continuity property of the wave number

$$k(z + h) = k(z) + hk'(z) + o(h) \quad \text{in} \quad x_0 < z < \infty. \tag{7}$$

Hence, we have the following wave equations in the different regions

$$\Psi''(z) + k^2(z)\Psi = 0; \quad x_0 < z < x \tag{8}$$

$$\Psi'' + k_0^2\Psi = 0 \qquad z < x_0 \tag{9}$$

$$\Psi'' + k_1^2\Psi = 0 \qquad z > x \tag{10}$$

We imagine a left moving wave at $z = \infty$ to travel down and impinge on the medium at $z = x_a$. This naturally produces a transmitted wave beyond x_0 and reflected wave in the regions $z > x$, with wave number corresponding to those regions. Therefore, we have

$$\Psi(z) = e^{-ik_1(z-x)} + \xi(x)e^{ik_1(z-x)}; \quad z > x \tag{11}$$

$$\Psi(z) = \tau(x)e^{-ik_0(z-x_0)}; \qquad z < x_0 \tag{12}$$

To arrive at the differential equations governing $\xi(x)$ and $\tau(x)$ the reflection and transmission functions, we apply the imbedding principle and extend the region (x_0 to x) by a small amount Δ at the end $z = x$ and assume that the extended region possesses the same continuity properties as stated earlier. Locally the incoming wave and the reflected wave near $z = x$ will appear as $e^{-ik(x)(z-x)}$ and $\xi(x)e^{+ik(x)(z-x)}$. Hence, we have, because of the continuity of the wave function and its derivative, the following equations which in fact define $\xi(x)$ in a sense.

$$\Psi(x) = 1 + \xi(x) \tag{13}$$

$$\Psi'(x) = -ik(x)[1 - \xi(x)]. \tag{14}$$

We will now extend the material at $z = x$ to $z = x + \Delta$ and consider the increase in the amount of the contributions to the reflection funtion $\xi(x)$.

The reflected flux from the medium of length $x + \Delta$ will be a sum of three parts taking into account the terms of order unity and first order in Δ:

(i) The immediate reflection from the stratum between x and $x + \Delta$. We take the wave number to be $k(x + \Delta)$ to the right of $z = x + \Delta$, and equal to $k(x)$ in the region to the right of x between x and $x + \Delta$.

(ii) The wave that arises from transmission through $(x + \Delta)$ to x, and reflection from the interface $z = x$, (from the subrod of length x) and again transmission through the interface $z = x + \Delta$.

(iii) The wave that results from the transmission through $x + \Delta$, reflection at x, turning back from $x + \Delta$, due to reflection and again refelction from the subrod of length x at the interface $z = x$ and finally transmission through

$z = x + \Delta$. The left moving incident wave at $z = x + \Delta$ gains in phase during the passage up and back in the region between $x + \Delta$ and x. This can be computed assuming the wave number $k(x)$ to be constant in this stratum. Hence, on the basis of the localization principle, we can write, taking into account the three types of events described above,

$$\xi(x+\Delta) = \frac{k(x+\Delta)-k(x)}{k(x+\Delta)+k(x)} + \frac{2k(x+\Delta)}{k(x+\Delta)+k(x)}\xi(x)e^{2ik(x)\Delta}\frac{2k(x)}{k(x+\Delta)+k(x)} + \tag{15}$$

$$+ \frac{2k(x+\Delta)}{k(x+\Delta)+k(x)}e^{ik(x)\Delta}\xi(x)e^{ik(x)\Delta}\frac{k(x)-k(x+\Delta)}{k(x)+k(x+\Delta)}e^{ik(x)\Delta}\left(\xi(x)e^{ik(x)\Delta}\frac{2k(x)}{k(x+\Delta)+k(x)}\right)$$

This can be reduced in view of Equation (7)

$$\xi(x+\Delta) = \xi(x)\left[1 + 2ik(x)\Delta - \xi(x)\frac{k'(x)\Delta}{2k(x)}\right] + \frac{k'(x)}{2k(x)} \tag{16}$$

and hence on going to the limit $\Delta \to 0$, we obtain

$$\xi'(x) = \frac{k'(x)}{2k(x)} + 2ik(x)\xi - \frac{k'(x)}{2k(x)}\xi^2(x). \tag{17}$$

Since a genuine discontinuity exists at $z = x_0$, we have for the initial condition for the above equation

$$\xi(x_0) = \frac{k(x_0+0)-k_0}{k(x_0+0)+k_0}. \tag{18}$$

To obtain the differential equation for $\tau(x)$ we argue as follows and add contributions to the transmitted flux terms up to first order in Δ.

(1) The incident beam at $z = x + \Delta$, gets transmitted through the stratum $[x + \Delta, x]$ and impinges on the subrod of length x, whose transmission function is known as $\tau(x)$.

(2) The incident flux goes through the initial region Δ, gets reflected at the interface $z = x$, from a subrod of length x, gets again reflected at the interface $z = x + \Delta$, goes back through Δ, and finally emerges at $z = x_0$

Taking these into account and adding the proper accumulation of phases we have

$$\tau(x+\Delta) = \frac{2k(x+\Delta)}{k(x+\Delta)+k(x)} e^{ik(x)\Delta}\left\{\tau(x)+\xi(x)e^{ik(x)\Delta}\left(\frac{k(x)-k(x+\Delta)}{k(x)+k(x+\Delta)}\right)e^{ik(x)\Delta}\tau(x)\right\}+0(\Delta). \tag{19}$$

As Δ tends to zero, we obtain the differential equation

$$\tau'(x) = \left[\frac{k'(x)}{2k(x)} + ik(x) - \frac{k'(x)}{2k(x)}\xi(x)\right]\tau(x) \tag{20}$$

with the initial condition corresponding to the discontinuity at $z = x_0$ given by

$$\tau(x_0) = \frac{2k(x_0 + 0)}{k(x_0 + 0) + k_0} . \tag{21}$$

Instead of the 'wavelet counting argument used above we can use analytic reasoning to obtain the differential Equations (17) and (20) for $\xi(x)$ and $\tau(x)$, respectively, if we consider the system

$$\Psi''(z) + k^2(z)\Psi(z) = 0; \quad x_0 \leq z \leq \infty \tag{22}$$

with the following continuity conditions defining $\tau(x)$ and $\xi(x)$

$$\Psi(x_0) = \tau(x); \quad \Psi'(x_0) = -ik_0\tau(x) \tag{23}$$

$$\Psi(x) = 1 + \xi(x); \quad \Psi'(x) = -ik(x)[1 - \xi(x)]; \qquad x_0 < x \tag{24}$$

where $k'(x)$ is at least piecewise continuous on $x_0 < z < \infty$.

Let $u_1(z)$ and $u_2(z)$ be two linearly independent solutions of (22) such that

$$\left.\begin{matrix} u_1(x_0) = 1 & u_2(x_0) = 0 \\ u_1'(x_0) = 0 & u_2'(x_0) = 1 \end{matrix}\right\} \tag{25}$$

Hence, the solution in the region $x_0 \leq z < \infty$ is given by

$$\Psi(z) = c_1u_1(z) + c_2u_2(z). \tag{26}$$

In view of the Equations (23), (24) and (25) we are led to the following

$$c_1 = \tau(x) \qquad c_2 = -ik_0\tau(x). \tag{27}$$

Hence

$$1 + \xi(x) = \tau(x)[u_1(x) - ik_0u_2(x)] \tag{28}$$

$$-ik(x)[1 - \xi(x)] = \tau(x)[u_1'(x) - ik_0u_2'(x)]. \tag{29}$$

Putting $\varphi = [u_1(x) - ik_0u_2(x)]$, we obtain by solving

$$\xi(x) = \frac{ik\varphi + \varphi'}{ik\varphi - \varphi'} \tag{30}$$

$$\tau(x) = \frac{2ik}{ik\varphi - \varphi'} \,. \tag{31}$$

Differentiation with respect to x yields the equation obtained earlier for $\xi(x)$ and $\tau(x)$ on the assumption that the denominators of (30) and (31) do not vanish.

$$\xi'(x) = 2ik(x)\xi(x) + \frac{k'(x)}{2k(x)}[1 - \xi^2(x)] \tag{32}$$

$$\tau'(x) = \left[\frac{k'}{2k} + ik - \frac{k'}{2k}\xi\right]\tau(x). \tag{33}$$

The above equations are the same as (17) and (20), respectively, with the initial conditions specified for those equations. With the aid of these functions we can easily obtain the internal solutions for the right-moving and left-moving wave inside the region of interest, as we did in the particle transport case. The different orders of backscattered solutions correspond to the terms of the Bremmer series solution [1] for the wave equation. Before illustrating this let us focus our attention on how Bremmer [1] derived the series solutions and the coupled integral equations for the two streams of the wave flux.

3. BREMMER SOLUTIONS

Let us imagine as we did earlier the three regions and the middle region between x_0 and x having a wave number $k(z)$ varying continuously with z, $x_0 \leq z \leq x$. However, let us push the end $z = x$ to ∞ and assume that the incident wave

$e^{ik_0(x_0-z)}$ impinges at the end $z = x_0$ which is exactly opposite to the situation considered earlier. We wish to be in conformity with the usual notation adopted in deriving Bremmer series solutions [1].

Following Bremmer, let us imagine the medium $x_0 \le z < \infty$ as being made up of layers $x_0 \le z < x_1$, $x_1 \le z \le x_2$, $x_2 \le z \le x_3$, etc., with constant wave numbers $k_1, k_2, \cdots, k_s$ in the layer $x_{s-1} \le z \le x_s$. Let us follow the beam that goes through all the layers and reach the point $z = (x_s - 0)$. Since we consider the beam as the result of only transmission through each layer up to x_s basing ourselves on the principle of localization, in view of the transmission coefficients pertinent to each interface Equation (6), summing up the phases gathered in each layer, we compute the transmitted flux as

$$u_0(x_s - 0) = \frac{2k_0}{k_0 + k_1} e^{ik_1\Delta x_1} \frac{2k_1}{k_1 + k_2} e^{ik_2\Delta x_2} \frac{2k_2}{k_2 + k_3} e^{ik_3\Delta x_3} \cdots \tag{34}$$

$$\cdots\cdots\cdots\cdots\cdots\cdots\cdots\cdots\cdots\cdots\cdots\cdots$$

$$= \exp\left\{ -\sum_{r=0}^{s-1} \log\left(1 + \frac{\Delta k_r}{2k_r}\right) + i \sum_{r=1}^{s} k_r \Delta x_r \right\}, \tag{35}$$

where $\Delta k_r = k_{r+1} - k_r$ is the difference of the wave numbers in succeeding layers. Passing to the limit as $\Delta x_r \to 0$, we easily see that (dropping the suffix r for x)

$$u_0(x) = \exp\left\{ -\tfrac{1}{2} \log \frac{k(x)}{k_0} + i \int_0^x k(r)\, dr \right\} \tag{36}$$

$$= \left[\frac{k_0}{k(x)}\right]^{\frac{1}{2}} \exp i \int_0^x k(r)\, dr. \tag{37}$$

Of course, Equation (37) is the JWKB solution for our particular initial condition, and this solution does not contain any contribution from any flux, arriving at the point x, due to reflections from any other point in the medium. Hence to supplement this we have to add reflected or backscattered waves arriving at the point z = x. To this end, we first compute the wave at x_r when the flux $u_0(x_r - 0)$ which has been entirely due to transmitted flux, undergoes reflection at this interface, giving rise to the left moving beam with one reflection with amplitude $v_1(x_r - 0)$

$$v_1(x_r - 0) = u_0(x_r - 0)\,\frac{k_r - k_{r+1}}{k_r + k_{r+1}}\,. \tag{38}$$

We now allow this wave to travel to the left to the layer just above the interface x_m by getting transmitted by the intervening layers. We can compute the ratio

$$\frac{v_1(x_m - 0)}{v_1(x_r - 0} = \exp\left\{-\sum_{\sigma=m+1}^{r} \log\left(1 - \frac{\Delta k_\sigma}{2k_\sigma}\right) + \sum_{\sigma=m+1}^{r} k_\sigma \Delta x_\sigma\right\}. \tag{39}$$

The transmission factor in Equation (39) is different from that in Equation (35) since the v_1 wave is traveling towards the left, i.e., opposite to the direction of the u_0 beam. The transmission factor at each interface $z = x_\sigma$ is $2k_\sigma/(k_{\sigma+1}+k_\sigma)$. Multiplying Equations (38) and (39) and going to the limit $\Delta x \to 0$ we obtain the wave $v_1(x)$ at x_m moving to the left having undergone one reflection at r given by

$$v_1(x_m) = -u_0(r)\,\frac{dk(r)}{2k(r)}\,\exp\left[\tfrac{1}{2}\int_{\sigma=x_m}^{\sigma=x_r} \frac{dk(\sigma)}{k(\sigma)} + i\int_{x_m}^{x_r} k(\sigma)d\sigma\right]. \tag{40}$$

All the layers for which $x_r > x_m$ will provide similar contributions to the first order reflected beam at x_m. Hence, we can integrate over r from x to ∞, and obtain* (dropping suffixes)

$$v_1(x) = \frac{-1}{2[k(x)]^{\frac{1}{2}}}\int_x^\infty dr\,\frac{k'(r)}{[k(r)]^{\frac{1}{2}}}\,u_0(r)\,\exp i\int_x^r k(\sigma)d\sigma. \tag{41}$$

---

* See H. Bremmer, Comm. on Pure and Appl. Math. 4 (1951) 105.

Now it is easy to imagine that the $v_1(x)$ beam at x by reflection will produce a right moving beam which in total has successively undergone two reflections. Hence, $u_2(x)$ will be coupled by an integral of the type (41) with $v_1(s)$ at all points s below x. This beam by another reflection will contribute to a v_3 beam moving left. Thus for original incidence at the left end, the right moving waves $u_{2N}(x)$ will be characterized by even numbers 2N(N = 0, 1, 2, ···) of reflections they have undergone until they reach the point of interest in the medium and the left moving waves $v_{2N+1}(x)$ would have undergone an odd number of reflections. The recurrence integral relations that will be satisfied by these two streams can be easily written down as

$$u_{2N}(x) = \delta_{N,0}\left\{\left[\frac{k_0}{k(x)}\right]^{\frac{1}{2}} \exp i \int_0^x k(s)\, ds\right\} + \left\{\frac{1}{2[k(x)]^{\frac{1}{2}}} \int_0^x ds\, \frac{k'(s)}{[k(s)]^{\frac{1}{2}}}\, v_{2N-1}(s) \exp i \int_s^x k(\sigma)\, d\sigma\right\}(1-\delta_{N,0}) \tag{42}$$

$$N = 0, 1, 2, \cdots, \infty$$

and

$$v_{2N+1}(x) = -\frac{1}{2[k(x)]^{\frac{1}{2}}} \int_x^\infty ds\, \frac{k'(s)}{[k(s)]^{\frac{1}{2}}}\, u_{2N}(s) \exp i \int_x^s k(\sigma)\, d\sigma \tag{43}$$

$$N = 0, 1, 2, \cdots, \infty$$

$\delta_{N,0}$ is the kronecker delta.

Of course, $u_0(x)$ will not be connected to the v flux and it is separately computed as $u_0(x)$ in Equation (37) and is the first term in the series for u(x) and provides the JWKB approximation for the solution of the original differential equation representing purely the transmitted flux up to the point x.

4. COUPLED DIFFERENTIAL AND INTEGRAL EQUATIONS FOR THE TWO BEAMS

We could obtain the above integral equations if we write down the differential equations for the $u(z)$ and $v(z)$, the right moving and left moving streams at any point z inside the medium $x_0 \leq z < \infty$ as we did in the case of the particle transport taking into account the reflection and transmission properties of the medium, for the two types of fluxes. We assume the principle of localization at each interface, and compute $u(z)$ from the u at $(z-\Delta)$ getting transmitted to z, and $v(z)$ from the flux v at $(z+\Delta)$ getting reflected at the interface z (Figure 1).

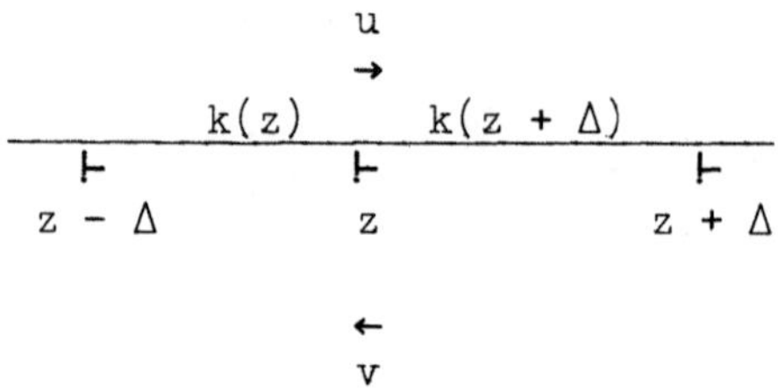

Figure 1

$$u(z + \Delta) = (1 - a\Delta)u(z) + b\Delta v(z + \Delta). \tag{44}$$

We assume that $(1 - a\Delta)$ is the fraction of $u(z)$ that gets through the interface at z when wave number changes from $k(z)$ to $k(z + \Delta)$. $b(z)\Delta$ is the fraction of $v(z)$ that turns over to the u flux by being reflected at the interface z, when trying to go from a layer of wave number $k(z + \Delta)$ to a layer of wave number $k(z)$. Hence, utilizing Equations (5) and (6), we find that

$$(1 - a(z)\Delta) = \frac{2k(z)}{k(z) + k(z + \Delta)} e^{ik\Delta} = \left\{1 - \left(\frac{k'}{2k} - ik\right)\Delta\right\} + o(\Delta) \tag{45}$$

and the reflection factor $b\Delta$ for the v beam is

$$b\Delta = \frac{k(z + \Delta) - k(z)}{k(z + \Delta) + k(z)} e^{ik\Delta} = \frac{k'\Delta}{2k} + o(\Delta). \tag{46}$$

Similarly, for the v(z) flux the forward equation is

$$v(z) = (1 - c\Delta)v(z + \Delta) + d\Delta u(z), \tag{47}$$

where $(1 - c\Delta)$ and $d\Delta$ are the fractions contributing to the v beam.

$$(1 - c\Delta) = \frac{2k(z + \Delta)}{k(z) + k(z + \Delta)} e^{ik\Delta} = \left\{1 + (ik + \frac{k'}{2k})\Delta\right\} + O(\Delta) \tag{48}$$

$$d\Delta = \frac{k(z) - k(z+\Delta)}{k(z) + k(z + \Delta)} e^{ik\Delta} = \frac{-k'}{2k}\Delta . \tag{49}$$

Starting from (44) and (47) and going to the limit as goes to zero, we obtain the coupled set of differential equations

$$\frac{du}{dz} = -\left(\frac{k'}{2k} - ik\right) u + \frac{k'}{2k} v \tag{50}$$

$$-\frac{dv}{dz} = \left(\frac{k'}{2k} + ik\right) v - \frac{k'}{2k} u. \tag{51}$$

We can have boundary conditions (for a medium extending from x_0 to x)

$$u(x_0) = u_0 \quad \text{and} \quad v(x) = 0. \tag{52}$$

Hence, if we label the u(z) and v(z) fluxes by the number of backscatterings they have undergone in the medium, until they reach the point z, by u_{2N}, v_{2N+1} we have the total u and v as

$$u(z) = \sum_{N=0}^{\infty} u_{2N}(z); \quad v(z) = \sum_{N=0}^{\infty} v_{2N+1}(z). \tag{53}$$

We can easily write down the transport equations for u_{2N} and v_{2N+1} in view of Equations (50) and (51). The first term in the right-hand side of Equations (50) and (51) represent no reflection, while the second coupling term involves one backscattering. Hence, we have (for the boundary condition (52)

$$\frac{du_{2N}}{dz} = -\left(\frac{k'}{2k} - ik\right) u_{2N} + \frac{k'}{2k} v_{2N-1} \tag{54}$$

$$\frac{dv_{2N+1}}{dz} = \frac{k'}{2k} u_{2N} - \left(\frac{k'}{2k} + ik\right) v_{2N+1} \tag{55}$$

The above equations can be obtained differentiating Equation (42) and (43) with respect to x. Hence, summing up the integral equations for all values of N, for u and v, we obtain at x = z

$$u(z) = u_0(z) + \frac{1}{2[k(z)]^{\frac{1}{2}}} \int_{x_0}^{z} \frac{k'(s)}{[k(s)]^{\frac{1}{2}}} v(s)\, ds \exp\left(i \int_{s}^{z} k(\sigma)\, d\sigma\right) \tag{56}$$

and

$$v(z) = \frac{-1}{2[k(z)]^{\frac{1}{2}}} \int_{z}^{\infty} ds \frac{k'(s)}{[k(s)]^{\frac{1}{2}}} u(s)\, ds \exp\left(i \int_{z}^{s} k(\sigma)\, d\sigma\right) \tag{57}$$

which can also be seen to be the result, if we integrate Equations (50) and (51) as they are.

Summing up Equations (50) and (51) we have for the total solutions $\Psi = u + v$ (which is the actual wave amplitude at any point) inside the medium the following relations

$$\begin{aligned} \Psi &= u + v \\ \Psi' &= u' + v' = ik(u - v). \end{aligned} \tag{58}$$

Hence, differentiating again with respect to z

$$\Psi'' = ik'(u - v) + ik[iku + ikv - \frac{k'}{k}u + \frac{k'}{k}v]. \tag{59}$$

By substituting for (u - v) from Euqations (50) and (51) and after a little algebra, we have

$$\Psi'' + k^2\Psi = 0. \tag{60}$$

This illustrates that u and v are the upgoing and downgoing solutions of the actual wave equation.

5. CONVERGENCE PROPERTIES OF THE SERIES SOLUTIONS

Since u and v satisfy the coupled integral Equations (56) and (57), the question naturally arises, whether the iterative solutions of (56) and (57) which are the well-known Bremmer series for u and v converge and, if so, under what conditions? These conditions can be stated as a theorem [2].

THEOREM 1. Let the function k(x) satisfy the conditions

$$\text{(a)} \qquad |k(x)| \geq a > 0 \quad x \geq x_0 \tag{61}$$

$$\text{(b)} \qquad \int_0^\infty |k'(x)|\, dx = b < 2a \tag{62}$$

Then the Bremmer series defined in Euqations (42), (43), and (53) converge to the solution of the wave equation (60).

In view of the condition a, the zero-th order term or the JWKB term of the series namely u_0 can be estimated as

$$|u_0(x)| = \frac{[k(x_0)]^{\frac{1}{2}}}{a^{1/2}} = q_1(\text{say}). \tag{63}$$

Hence, v_1 expressed by Equation (41) is bounded by

$$|v_1(x)| \leq \frac{q_1}{2a^{\frac{1}{2}}} \int_{x_0}^\infty \frac{k'(s)}{a^{1/2}} \leq \frac{q_1}{2a} \int_{x_0}^\infty |k'(s)|\, ds = \frac{q_1 b}{2a}\,. \tag{64}$$

We can show inductively that for some constant q the successive estimates of u_{2N} and v_{2N+1} are bounded:

$$|u_{2N}(x)| \leq q\left(\frac{b}{2a}\right)^{2N} \tag{65}$$

and

$$|v_{2N+1}(x)| \leq q\left(\frac{b}{2a}\right)^{2N+1}. \tag{66}$$

The above are known to be valid for N = 0.

Since it is easy to see from Equations (42) and (43) that

$$|u_{2N}(x)| \leq \frac{1}{2a^{\frac{1}{2}}} \int_{x_0}^{x} \frac{k'(s)}{a^{\frac{1}{2}}} q(\frac{b}{2a})^{2N-1} ds$$

$$\leq \frac{1}{2a} q(\frac{b}{2a})^{2N-1} \int_{x_0}^{\infty} |k'(s)| ds = q(\frac{b}{2a})^{2N}.$$

$$|v_{2N+1}(x)| \leq \frac{1}{2a^{\frac{1}{2}}} \int_{x}^{\infty} \frac{|k'(s)| ds}{a^{\frac{1}{2}}} q(\frac{b}{2a})^{2N} \leq (\frac{q}{2a})(\frac{b}{2a})^{2N} \int_{x_0}^{\infty} k'(s) ds$$

$$= q(\frac{b}{2a})^{2N+1}. \tag{67}$$

It follows that the Bremmer series converges uniformly for all $x \geq 0$ and that $\Psi = u + v$ is uniformly bounded for $x \geq 0$. However, the condition (b) of Equation (62) may be relaxed and convergence may still be established. From Equations (50), (51), and (58), we find that

$$u = \frac{\Psi}{2} + \frac{\Psi'}{2ik} \tag{68}$$

$$v = \frac{\Psi}{2} - \frac{\Psi'}{2ik} . \tag{69}$$

The influence of the inhomogeneity of the medium on the applicability of the JWKB approximation can be shown very clearly, if by exploiting Equations (50) and (51), we write down

$$\frac{d}{dx}\left[k^{\frac{1}{2}} \exp\left\{-i \int_{x_0}^{x} k(s) ds\right\} u\right] = \frac{k'}{2k} \left[k^{\frac{1}{2}} \exp\left\{-i \int_{x_0}^{x} k(s) ds\right\} v\right] \tag{70}$$

and

$$\frac{d}{dx}\left[k^{\frac{1}{2}} \exp\left\{ i \int_{x_0}^{x} k(s)\, ds\right\} v\right] = \frac{k'}{2k}\left[k^{\frac{1}{2}} \exp\left\{ i \int_{x_0}^{x} k(s)\, ds\right\} u\right]. \quad (71)$$

To obtain the above, we multiply Equation (56) by $[k(x)]^{\frac{1}{2}} \exp\left\{-i\int_{x_0}^{x} k(t)\, dt\right\}$ on both sides and differentiate both sides with respect to x. Similarly, we multiply both sides of Equation (57) by $k^{\frac{1}{2}} \exp\left\{ i\int_{x_0}^{x} k(t)\, dt\right\}$ and differentiate both sides. In view of Equation (37) for u_0, we have

$$\frac{d}{dx}\left[u_0 k^{\frac{1}{2}} \exp\left\{ -i \int_{x_0}^{x} k(t)\, dt\right\}\right] = 0. \quad (72)$$

This means that for small values of k'/2k the right hand sides of Equations (70) and (71) can be neglected and we have the JWKB approximation for the solution Ψ.

Having shown the convergence of the Bremmer series solution (under certain conditions), it may be instructive to see that one may also demonstrate that the pair of integral equations, (56) and (57), represent the original wave equation.

Carrying out the differentiation in (70) and (71), we have

$$u'(x)k^{\frac{1}{2}}(x) + \frac{1}{2}\frac{k'}{[k]^{\frac{1}{2}}} u(x) - ik^{3/2}(x)u(x) = \frac{1}{2}\frac{k'(x)}{[k(x)]^{\frac{1}{2}}} v(x) \quad (73)$$

$$v'(x)k^{\frac{1}{2}}(x) + \frac{1}{2}\frac{k'}{[k]^{\frac{1}{2}}} v(x) + ik^{3/2}(x)v(x) = \frac{1}{2}\frac{k'(x)}{[k(x)]^{\frac{1}{2}}} u(x). \quad (74)$$

Adding the above two equations

$$[u'(x) + v'(x)] + ik(x)[v(x) - u(x)] = 0. \quad (75)$$

Substracting Equations (74) from (73), we obtain

$$k(u' - v') + k'(u - v) - ik^2(u + v) = 0. \quad (76)$$

Differentiating Equation (75) with respect to x, we have

$$[u'' + v''] + ik'[v - u] + ik[v' - u'] = 0. \tag{77}$$

Multiplying Equation (77) by i, and subtracting (76) we arrive at

$$u'' + v'' + k^2(u + v) = 0 \tag{78}$$

or equivalently

$$\Psi'' + k^2(x)\Psi = 0. \tag{79}$$

It was stated in [2] by Bellman and Kalaba, regarding the conditions for the convergence of the Bremmer series, a weaker condition like

$$|k(x)| \geq a > 0 \text{ and } \int_0^\infty |k'(x)|dx < \infty \tag{80}$$

may not be enough. They assumed another condition

$$\int_0^\infty |k'(x)| \; dx < 2a. \tag{81}$$

This was brought out rigorously by F.V. Atkinson [5] who showed that the series converges if

$$\int_{x_0}^\infty |k'(x)/k(x)|dx \leq \Pi \tag{82}$$

and that the number Π cannot be replaced by any larger number. To demonstrate this statement, we first observe that the JWKB solutions of the u and v fluxes, which can be obtained from Equation (50) and (51) by neglecting the non-diagonal parts, can be written down as

$$u_0(x, s) = y(x, s) = y(x)[y(s)]^{-1}, \tag{83}$$

where

$$y(x) = [k(x)]^{-1/2} \exp\left(i \int_0^x k(\tau) \; d\tau\right). \tag{84}$$

This u_0 represents the forward transmitted wave at x with unit amplitude at s. It ignores the contributions from the backscattering in the region from s to x. Similarly, the complex conjugate of y(x, s) which we may call z(x, s) is the transmission of the wave moving in the opposite direction, i.e., the v flux. The secondary wave set up at each point by two fluxes depends on the local reflectivity, which is $-\frac{1}{2}\frac{k'}{k}$ for the u flux and $+\frac{1}{2}\frac{k'}{k}$ for the v flux as explained earlier. Hence, we can rewrite Equations (56) and (57) in a general form as

$$u(x) = y(x, 0) + \int_0^x y(x, s)[\tfrac{1}{2} k^{-1}(s)k'(s)]v(s)\, ds \tag{85}$$

$$v(x) = \int_x^\infty z(x, s)[-\tfrac{1}{2} k^{-1}(s)k'(s)]u(s)\, ds. \tag{86}$$

We assume the condition that

$$\int_0^\infty |k'/k|\, dx < \infty \tag{87}$$

and if k(x), for $x \geq 0$ be positive and continuously differentiable, the integral equations, have a unique solution in the class of functions which are continuous and uniformly bounded in $x \geq 0$. Equation (87) implies that both k'(x) and $k^{-1}(x)$ are uniformly bounded in x, so that y(x, s) and z(x, s) are likewise uniformly bounded. Then $\Psi = u + v$ is the solution of the second order differential equation. This has already been illustrated by Equations (73) to (79).

To substantiate the statement that the pair of integral equations (85) and (86) has a unique solution, appeal is made to the idea of the Fredholm alternative, according to which the equation is uniquely solvable, provided the homogeneous equation has no solution other than the trivial solution which is the zero solution.

We consider the set of Equations (85) and (86) as a linear equation in Banach space R of pairs of functions {u(x), v(x)} both of which are to be continuous and uniformly bounded for $x \geq 0$ the norm being

$$\ell.\ u.\ b\ \max\{|u(x)|, |v(x)|\}; \tag{88}$$

$$0 \leq x < \infty$$

The integral operators in (85) and (86) are clearly bounded in this norm and, because of the nature of the functions y(x, s) and z(x, s) and in view of the condition (87), these integral operators are shown to be completely continuous in R [6]. According to Riesz theory [6] of completely continuous operators, the set (85) and (86) is uniquely solvable, unless there is a pair {u(x), v(x)} $\in$ R satisfying, (85) and (86) with y(x) replaced by zero and not both identically zero. This means that we have to consider the following set of simultaneous eigenvalue equations,

$$u(x) = \lambda \int_0^x y(x, s)[\tfrac{1}{2}k^{-1}(s)k'(s)]v(s)\, ds \tag{89}$$

$$v(x) = \lambda \int_x^\infty z(x, s)[-\tfrac{1}{2}k'(s)k^{-1}(s)]u(s)\, ds \tag{90}$$

and prove that $\lambda = 1$ is not an eigenvalue. However, it is proved in [5] that the above homogeneous set of equations have no real eigenvalues at all. To this end we convert (89) and (90) into the following differential equations along with the boundary conditions:

$$u' = (ik - \frac{k'}{2k})u + \frac{1}{2}\frac{\lambda k'}{k} v \tag{91}$$

$$v' = (-ik - \frac{k'}{2k})v + \frac{1}{2}\frac{\lambda k'}{k} u \tag{92}$$

$$u(0) = 0 \qquad v(\infty) = 0. \tag{93}$$

If λ is real, the conservation of current flow prescribes the following condition for the solutions u and v of the set (91) and (92)

$$k|u|^2 - k|v|^2 = \text{constant}. \tag{94}$$

This can be proved easily bi differentiating $(k|u|^2)$ and $(k|v|^2)$ to obtain

$$(k|u|^2)' = \frac{1}{2}\{ \bar{\lambda}\bar{v}k'u + \lambda\bar{u}k'v\} \tag{95}$$

and

$$(k|v|^2)' = \frac{1}{2}\{\bar{\lambda}\bar{u}k'v + \lambda\bar{v}k'u\}, \tag{96}$$

where $\bar{\lambda}$, $\bar{u}$ and $\bar{v}$ are the complex conjugates of λ, u and v. If λ is real and $\bar{\lambda} = \lambda$ by subtracting (96) from (95), we obtain

$$(k|u|^2 - k|v|^2)' = 0. \tag{97}$$

Applying the conservation condition at x = 0 and x = ∞, we have in view of the homogeneous set of Equations (89) and (90)

$$-k|v(0)|^2 = k|u(x)|^2_{x \to \infty}. \tag{98}$$

The left hand side is negative while the right hand side is non-negative. Hence, v(0) = 0. This means both u(0) and v(0) are zero, which leads to the conclusion that u(x) and v(x) are identically zero for all $x \geq 0$. Hence, the homogeneous set (90) and (91) does not possess nontrivial solutions for any real value of λ, i.e., any real value of λ is not an eigenvalue of the set.

Let us now introduce the parameter λ in the coupled set of inhomogeneous integral equations (85) and (86) to obtain the system

$$u(x) = y(x) + \frac{1}{2}\lambda \int_0^x y(x, s)k'(s)k^{-1}(s)v(s)\, ds \tag{99}$$

$$v(x) = -\frac{1}{2}\lambda \int_x^\infty z(x, s)k^{-1}(s)k'(s)u(s)\, ds. \tag{100}$$

This set has the well-known Neumann series as their solutions, i.e., a power series in the parameter λ, for sufficiently small values of λ.

$$u(x) = u_0(x) + \lambda^2 u_2(x) + \lambda^4 u_4(x) + \cdots \tag{101}$$

$$v(x) = \lambda v_1(x) + \lambda^3 v_3(x) + \lambda^5 v_5(x) \cdots, \tag{102}$$

where u_{2N} and v_{2N+1} are successive approximations, and the total solution $\Psi = \Sigma u_{2N} + \Sigma v_{2N+1}$. The above series converge absolutely if $|\lambda| < |\lambda_1|$ where λ_1 is the smallest in absolute value of the eigenvalues of the homogeneous set of equations, (89) and (90). The convergence being in R will also be uniform. This means that for the Bremmer series which represents the solution of the set (85) and (86) to converge, the least eigenvalue λ_1 satisfies the condition

$$|\lambda_1| > 1. \tag{103}$$

To realize this condition the main criterion was stated and proved as a theorem in Reference [5].

Let k(x), $x \geq 0$ be positive continuously differentiable and let

$$\int_0^\infty |k^{-1}(x)k'(x)|dx \leq \Pi. \tag{104}$$

Then the Bremmer series is absolutely and uniformly convergent.

Since the homogeneous set of equations, (89) and (90), correspond to the set of differential equations (91) and (92), with boundary conditions (93), we have to show that for k(x) being a continuously differentiable function and positive for all $x \geq 0$ and satisfying Equation (87) the system (89) and (90) has no eigenvalues satisfying

$$|\lambda|\int |k^{-1}(x)k'(x)|dx \leq \Pi. \tag{105}$$

To this end we introduce polar coordinates and express u and v as

$$u(x) = r_1 \exp i\,\theta_1 \quad \text{and} \quad v(x) = r_2 \exp i\,\theta_2 \tag{106}$$

with the boundary conditions

$$r_1(0) = 0 \quad \text{and} \quad r_2(\infty) = 0. \tag{107}$$

In order that the system has no nontrivial solutions, the condition

$$r_2(0) \neq 0 \quad \text{and} \quad r_1(\infty) \neq 0 \tag{108}$$

should hold.

Since $[k^{-1}(x)k'(x)]$ is absolutely integrable over $[0, \infty]$, a result of Levinson (p. 50, Reference [7]) enables us to write the solution of the set (91) and (92) in the following form as $x \to \infty$,

$$u \sim \exp i\int_0^x k(s)\,ds[c_1 + o(1)] \tag{109}$$

$$v \sim \exp -i \int_0^x k(s)\, ds[c_2 + o(1)], \tag{110}$$

where c_1 and c_2 are constants, which can vanish simultaneously only for the trivial solution.

Hence, from conditions (107) and (108) we find that the ratio r_1/r_2 increases from 0 to ∞, as x increases from 0 to ∞ or over a subinterval of $(0, \infty)$. Hence, it is necessary to conclude that $\tan^{-1}(r_1/r_2)$ should vary at least by $\Pi/2$ as x goes from 0 to ∞. Hence, we must have the condition

$$\int_0^\infty \left|\frac{d}{dx} \tan^{-1} \frac{r_1}{r_2}\right| dx \geq \frac{1}{2}\Pi. \tag{111}$$

To evaluate the integrand, we see that

$$\frac{d}{dx} \tan^{-1} \frac{r_1}{r_2} = \frac{r_2^2}{r_1^2 + r_2^2}\left(\frac{r_1}{r_2}\right)' \tag{112}$$

where $r_1 = \sqrt{u\,\bar{u}}$ and $r_2 = \sqrt{v\,\bar{v}}$. However, we can express

$$\frac{r_1}{r_2} = \sqrt{\frac{\bar{u}ku}{\bar{v}kv}} = \sqrt{\frac{f}{g}}. \tag{113}$$

We exploit the relations (95) and (96) and write

$$(\bar{u}ku)' = \mathrm{Re}\{\lambda\bar{u}k'v\}; \quad (\bar{v}kv)' = \mathrm{Re}\{\lambda\bar{v}k'u\}. \tag{114}$$

If we now express λ in the polar form

$$\lambda = \mu \exp i\varphi \tag{115}$$

we have

$$(\bar{u}ku)' = f' = \mu r_1 r_2 \cos(\varphi + \theta_2 - \theta_1)k' \tag{116}$$

$$(\bar{v}kv)' = g' = \mu r_1 r_2 \cos(\varphi + \theta_1 - \theta_2)k'. \tag{117}$$

Thus, the integrand in (111) becomes

$$\frac{d}{dx}\tan^{-1}\frac{r_1}{r_2} = \frac{\mu k'}{2k}\left\{\frac{r_2^2\cos(\varphi+\theta_2-\theta_1)-r_1^2\cos(\varphi+\theta_1-\theta_2)}{r_1^2+r_2^2}\right\}. \tag{118}$$

Thus, we easily deduce

$$\left|\frac{d}{dx}\tan^{-1}\frac{r_1}{r_2}\right| \le \left|\frac{\mu k'}{2k}\right| \tag{119}$$

and since $\mu = |\lambda|$

$$|\lambda| \int_0^\infty \left|\frac{k'}{k}\right| dx \ge \Pi. \tag{120}$$

Thus, Equation (120) has proved the theorem, that no eigenvalues for the homogeneous set (89) and (90) satisfy the condition expressed in Equation (105) except for the case implied by the equality sign. From Equation (119), if the equality

$$\int_0^\infty\left|\frac{d}{dx}\tan^{-1}\frac{r_1}{r_2}\right|dx = \int_0^\infty\left|\frac{\mu k'}{2k}\right|dx = \frac{\Pi}{2} \tag{121}$$

is assumed to hold, we obtain the following situation for all $x \geq 0$. Either

$$\cos(\varphi+\theta_2-\theta_1) = 1 \quad\text{and}\quad \cos(\varphi+\theta_1-\theta_2) = -1 \tag{122}$$

or

$$\cos(\varphi+\theta_2-\theta_1) = -1 \quad\text{and}\quad \cos(\varphi+\theta_1-\theta_2) = 1 \tag{123}$$

with the possible exception of these values of x, for which $k'(x) = 0$. This means that both φ and $\theta_2 - \theta_1$ are both odd multiples of $\Pi/2$ except possibly when $k'(x) = 0$. This means that λ as well as $\bar{v}u$ and $(\bar{u}v)$ are <u>pure imaginary</u> except possibly when $k' = 0$. We now prove that when $k' \neq 0$, the equality (121) is not possible. To this end we deduce from Equations (91), (92) and (93)

$$(\bar{u}kv)' = -2i\bar{u}k^2v + \frac{1}{2}\bar{\lambda}\bar{v}k'v + \frac{1}{2}\lambda\bar{u}k'u. \tag{124}$$

We know that $(\bar{u}kv)'$ is pure imaginary, so are the terms $(\frac{1}{2}\lambda\bar{v}k'v + \frac{1}{2}\bar{u}k'u)$. Hence in the interval in which $k' \neq 0$ the first term which is real has to vanish, i.e., $\bar{u}k^2v = 0$ throughout the interval in which $k' \neq 0$. Since u and v have only isolated zeroes, this is not possible. Hence there is no interval in which $k' \neq 0$. This means k' has to be zero throughout the interval which makes the equality

$$\int \frac{d}{dx}\left|\tan^{-1}\frac{r_1}{r_2}\right|dx = \frac{1}{2}\Pi = \int \left|\frac{k'}{2k}\right| dx \quad \text{untenable.} \tag{125}$$

Thus the equality in (121) is impossible for any eigenvalues and hence, there are no eigenvalues for the homogeneous set (89) and (90) of the integral equations, satisfying Equation (105). Hence the Bremmer series uniformly and absolutely converges for the stated condtion (104) only.

For alternate methods of establishing the convergence of the series, the reader is referred to the article of G.M. Wing [11].

6. BREMMER SERIES USING FINITE ORDER SCATTERING REFLECTION AND TRANSMISSION FUNCTIONS

In Section 1 of this chapter, we have arrived at the equations for the reflected and transmitted beams $\rho(x)$* and $\tau(x)$ for the wave propagation for the given boundary conditions. (Equations (17) and (20).) We have also seen how wave propagation can be represented by a set of transport equations for u and v, the right-going and left-going beams as in Equations (50) and (51). For the particle transport too we obtained a coupled set of equations for u and v and wrote down equations for the reflection and transmission functions for incidence at the right end and left end. We also analyzed in Chapter III the equations governing the emergent fluxes which have undergone backscatterings inside the medium, a finite number of times using the imbedding method. Using these finite order-of-scattering reflection and transmission functions we consider here transport problems relating to wave propagation, as represented by Equations (50) and (51). We can arrive at the JWKB approximation as well as the Bremmer series for the solution of the wave equation. These ideas have been dealt with in References [10] and [11]. Let us write down in a general form the equation governing the transport of the u and v beams.

---

* We designated these functions as $\xi(x)$ and $\tau(x)$ in Section (1) of this Chapter. From now on we use the notation $\rho(x)$ and $\tau(x)$ for reflection and transmission functions with input at x i.e. with $v(x) = 1$ and $u(0) = 0$.

$$\frac{du}{dx} = A(z)u(z) + B(z)v(z) \qquad (126)$$

$$y \leq z \leq x$$

$$\frac{-dv}{dz} = C(z)u(z) + D(z)v(z) \qquad (127)$$

with the coefficients A, B, C and D given by the corresponding expressions in the set (50) and (51). The boundary conditions can be either

$$u(y) = 0, \quad v(x) = 1 \qquad (128)$$

or

$$u(y) = 1 \text{ and } v(x) = 0. \qquad (129)$$

We can choose them according to the needs of the problem. Herein let us agree to call $\rho(x)$ and $\tau(x)$ the reflection and transmission functions for the v beam corresponding to the boundary condition (128). Let us write generally $\rho(y, x)$ the reflection function for the v beam, i.e., with the unit input at the end x, and no incidence at the end $z = y$. Similarly $\tau(y, x)$ is the transmission for the v beam with input at the end $z = x$, and no incidence at $z = y$.

$$\rho(y, x) = u(x) \text{ and } \tau(y, x) = v(y). \qquad (130a)$$

We now introduce $R(y, x)$ and $T(y, x)$ for the u beam corresponding to the boundary conditions (129). In other words, $R(y, x)$ is the reflected flux at y, for unit incidence at y, and no input at x. Similarly, $T(y, x)$ is the transmitted flux at $z = x$ due to unit incident flux at $z = y$, and no incidence at $z = x$.

$$R(y, x) = v(y) \text{ and } T(y, x) = u(x). \qquad (130b)$$

We have assumed that for all y, x, and z, such that for $Y \leq y \leq z \leq x \leq X$ solutions do exist.

Employing the Riccati transformations as elaborated in Chater III, we can arrive at the following set of differential equations for ρ, τ, R, and T.

$$\frac{d\rho(y, x)}{dx} = B + (A + D)\rho + C\rho^2 \qquad (131)$$

$$\frac{d\tau(y, x)}{dx} = [D(x) + C(x)\rho(y, x)]\tau(y, x) \qquad (132)$$

$$\frac{dR(y, x)}{dy} = -C(y) - R(y, x)[A(y) + D(y)] - R^2(y, x)B(y) \tag{133}$$

$$\frac{dT(y, x)}{dx} = [A(x) + \rho(y, x)C(x)]T(y, x) \tag{134}$$

$$\frac{d\tau(y, x)}{dy} = -[D(y) + R(y, x)B(y)]\tau(y, x) \tag{135}$$

$$\frac{d\rho(y, x)}{dy} = -T(y; x)B(y)\tau(y; x). \tag{136}$$

$$\frac{dR(y, x)}{dx} = \tau(y, x)C(x)T(y, x) \tag{137}$$

$$Y \le y \le x \le X$$

$$\frac{dT(y, x)}{dy} = -T(y, x)\left[A(y) + B(y)R(y, x)\right] . \tag{138}$$

These equations can be derived by the use of Riccati transformations and the corresponding recovery transformations [8]. The variations of the medium can take place at either the incident end or at the farther end. These can also be derived by the particle counting techniques. Also equations for finite order of scattering functions, ρ_{2N+1}, τ_{2N}, R_{2N+1}, and T_{2N}, can be easily obtained by using physical arguments (by employing input-output analysis in a segment for each of these functions). They represent a mode of linearization of the above set (131 to 138) [9].

The initial values of the set (131) to (138) are:

$$R(\xi, \xi) = 0 = \rho(\xi, \xi) \qquad \tau(\xi, \xi) = T(\xi, \xi) = 1 \tag{139}$$

$$Y \le \xi \le X.$$

If we concentrate on the boundary conditions, Equation (139), we arrive at the JWKB solutions by solving Equation (134) ignoring reflections completely.

$$\frac{dT_0(y, z)}{dz} = AT_0. \tag{140}$$

In the above we have assumed $\rho = 0$, in Equation (134) and the contributions to the u beam from backscatterings are completely lost sight of. In view of the initial conditions for T, we obtain

$$T_0(y, z) = \exp \int_y^z A(t)\, dt. \tag{141}$$

By definition the transmitted flux u(z) is

$$u_0(z) = T_0(y, z) u_0(y). \tag{142}$$

If the forward transmission factor A, in each Δ, for wave propagation is expressed as in Equations (50) and (51),

$$u_0 = \Psi_0(z) \sim u_0(y) \left[\frac{k(y)}{k(z)}\right]^{\frac{1}{2}} \exp i \int_y^z k(t)\, dt. \tag{143}$$

The above is the first term of the Bremmer series or the JWKB approximation for the given boundary conditions.

If we now want to find the $v_1(z)$ beam that has been created by one backscattering of u beam, we take into consideration the linearized set of equations corresponding to Equation (133) as in [9].

$$\frac{dR_{2N+1}(y, x)}{dy} = -C\delta_{N,\,0} - R_{2N+1}(A + D) - \sum_{m=0}^{N-1} R_{2m+1} B \times R_{2(N-m)-1} \tag{144}$$

For N = 0,

$$\frac{dR_1(z, x)}{dz} = -C(z) - R_1(A + D). \tag{145}$$

The solution for $R_1(z, x)$ is

$$R_1(z, x) = - \int_x^z ds C(s) \exp \int_z^s D(t)\, dt \exp \int_z^s A(t')\, dt'. \tag{146}$$

We know:

$$\exp \int_z^s A(t)\, dt = \frac{u_0(s)}{u_0(z)}. \tag{147}$$

Since it can be easily seen that

$$u_0(z)R_1(z, x) = v_1(z, x) = - \int_x^z C(s)u_0(s)\, ds \exp \int_z^s D(t)\, dt \qquad (148)$$

we obtain

$$v_1(z, x) = - \frac{1}{2[k(z)]^{\frac{1}{2}}} \int_z^x \frac{k'(s)}{[k(s)]^{\frac{1}{2}}} u_0(s)\, ds \exp i \int_z^s k(t)\, dt. \qquad (149)$$

We could in Equation (133) replace the terms $-R(D + RB)$ making use of Equation (135) and obtain

$$\frac{dR(z, x)}{dz} = -C - RA + R \frac{d\tau(z, x)}{dz} (\tau(z, x))^{-1}. \qquad (150)$$

In the above if we substitute $\tau_0(z, x)$ for $\tau(z, x)$, Equation (150) represents the differential equation for $R_1(z, x)$. We can then obtain the same result as (149) for the once scattered beam $v_1(z, x)$. Of course, a direct analysis of Equation (150) will lead to

$$v_1(z, x) = -\int_x^z C(s)u_0(s)\, ds \exp \int_z^s D(t)\, dt. \qquad (151)$$

However, if we want to obtain the beam $u_2(z, x)$ which has experienced two backscatterings, we should refer back to the equation for the internal intensities inside the medium in terms of the reflection and transmission functions. We have seen in Chapter III that

$$u(z, x) = T(y, z)u_0(y) + v(z, x)\rho(y, z) \qquad (152)$$

$$v(z, x) = R(z, x)u(z, x). \qquad (153)$$

These hold good for incidence $u_0(y)$ at the end y and no input at the end $z = x$.

For the finite order of scattering solutions we have

$$u_{2N}(z, x) = T_{2N}(y, x)u_0(y) + \sum_{m=0}^{N-1} v_{2m+1}\rho_{2(N-m)-1} \qquad (154)$$

$$v_{2N+1}(z, x) = \sum_{m=0}^{N} R_{2(N-m)+1}\, u_{2m}(z, x). \qquad (155)$$

Hence, to compute twice scattered $u_2(z; x)$ we have

$$u_2(z, x) = T_2(y, z)u_0(y) + v_1(z, x)\rho_1(y, z). \qquad (156)$$

Let us compute the function $T_2(y, z)$ from Equation (134), which can be decomposed as an infinite set of differential equations.

$$\frac{dT_{2N}(y, z)}{dz} = T_{2N}A + \sum_{m=0}^{N-1} \rho_{2(N-m)-1}CT_{2m} \qquad (157)$$

with the initial condition

$$T_{2N}(y, y) = \delta_{N,\,0}. \qquad (158)$$

Hence, we have

$$\frac{dT_2(y, z)}{dz} = T_2(y, z)A(z) + \rho_1(y, z)C(z)T_0(y, z). \qquad (159)$$

$T_2(y, z)$ can be found as

$$T_2(y, z) = \int_y^z C(s)\rho_1(y, s)\, ds \left(\exp \int_s^z A(t)\, dt\right) T_0(y, s). \qquad (160)$$

The equation governing $\rho_1(y, z)$ can be found from the linearizing set corresponding to Equation (131)

$$\frac{d\rho_{2N+1}(y, z)}{dz} = B\delta_{N,\,0} + (A + D)\rho_{2N+1} + \sum_{m=0}^{N-1} \rho_{2m+1}C\rho_{2(N-m)-1} \qquad (161)$$

with $\rho_{2N+1}(y, y) = 0$ for all N.

Hence

$$\frac{d\rho_1(y, z)}{dz} = B + (A + D)\rho_1 \qquad (162)$$

and

$$\rho_1(y, z) = \int_y^z B(\ell)\, d\ell \exp \int_\ell^z (A + D)\, dt. \tag{163}$$

From Equations (160) and (163) we find

$$u_0(y)T_2(y, z) = \int_y^z C(s)u_0(s)\, ds \int_y^s d\ell B(\ell) \exp\int_\ell^s (A+D)(t)\, dt \exp\int_s^z A(t')\, dt'. \tag{164}$$

Changing the order of integration between ℓ and s, we obtain

$$u_0(y)T_2(y, z) = \int_y^z d\ell B(\ell) \exp \int_\ell^z A(t)\, dt \int_\ell^z ds C(s)u_0(s) \exp\int_\ell^s D(t')\, dt'. \tag{165}$$

We express the second term in Equation (156) in view of the expression (151) for $v_1(z, x)$ and (163) for ρ_1 as

$$\rho_1(y, z)v_1(z, x) = \int_y^z B(\ell)\, d\ell \exp \int_\ell^z (A+D)\, dt \int_z^x C(s)u_0(s)\, ds \exp\int_z^s D(t')\, dt' \tag{166}$$

$$= \int_y^z B(\ell)\, d\ell \exp \int_\ell^z A(t)\, dt \int_z^x C(s)u_0(s)\, ds \exp \int_\ell^s D(t')\, dt'.$$

Hence, adding (165) and (166) we obtain

$$u_2(z, x) = \int_y^z B(\ell)\, d\ell \exp \int_\ell^z A(t)\, dt \int_\ell^x C(s)u_0(s)\, ds \exp \int_\ell^s D(t')\, dt'$$

$$= \int_y^z B(\ell)v_1(\ell, x) \exp \left[\int_\ell^z A(t)\, dt\right] d\ell \tag{167}$$

$$= \frac{1}{2\sqrt{k(z)}} \int_y^z \frac{k'(\ell)}{\sqrt{k(\ell)}} v_1(\ell, x) d\ell \exp\left(i \int_\ell^z k(t')\, dt'\right).$$

The next entity to be computed is $v_3(z, x)$, and it can be shown using the relation (refer to Equation (155))

$$v_3(z, x) = R_3(z, x)u_0(z, x) + R_1(z, x)u_2(z, x). \tag{168}$$

Where R_3 can be explicitly found from the differential equation (see Equation (144))

$$\frac{dR_3(z, x)}{dz} = -R_3(z, x)[A(z) + D(z)] - R_1(z, x)B(z)R_1(z, x). \tag{169}$$

The computation yields

$$v_3(z, x) = \int_z^x C(s)u_2(s)\ ds\ \exp \int_z^s D(t)\ dt. \tag{170}$$

Thus we have the iteration scheme derived for the Bremmer series which was derived earlier directly from the differential equations for u_{2N} and v_{2N+1}. Here we have obtained the internal order of scattering solutions by utilizing the reflection and transmission functions of different orders.

7. WAVE EQUATIONS WITH A SOURCE TERM

Let us now consider the wave equation

$$\Psi'' + k^2(z)\Psi = f(z) \quad Y \le z \le X \tag{171}$$

with the same boundary conditions as mentioned earlier. How can we go about writing down the transport equations corresponding to this problem? Let us argue in an intuitive fashion and arrive at the set of first order equations similar to (50) and (51) in this section. A regular procedure to handle this situation will be described in Chapter VI.

We know that in solving the second order equation of the type (171), we write down the particular solution using the Green's function $G(z, z')$. This expresses the effective amplitude produced at the point z, due to a source of unit amplitude at z'. Hence, we integrate z' over the entire interval to express the effect of all the sources, and the particular integral becomes $\int G(z, z')f(z')\ dz'$. We assume the principle of localization and take k to be constant inside each stratum, Δ. The amount of wave amplitude produced at the ends of the stratum Δ, between z and z + Δ, may be taken to be $(e^{\pm ik\Delta}/2ik)f(z)\Delta$, for free boundary conditions at the ends of the thin region around z. The + sign refers to the upgoing amplitude and te - sign to the amplitude in the other direction (refer to Morse and Freshbach [12]). Arguing in this heuristic manner we arrive at the transport equations for the upgoing and downgoing wave as

$$\frac{du}{dz} = (ik - \frac{k'}{2k})u + \frac{k'}{2k}v + \frac{f(z)}{2ik} \tag{172}$$

$$\frac{dv}{dz} = -(ik + \frac{k'}{2k})v + \frac{k'}{2k}u - \frac{f(z)}{2ik}. \tag{173}$$

Since the total solution $\Psi = u + v$, we have

$$\frac{d\Psi}{dz} = ik(u - v). \tag{174}$$

Also we find subtracting (173) from (172)

$$\frac{d}{dz}(u - v) = ik(u + v) + \frac{k'}{k}(v - u) + \frac{f(z)}{ik} . \tag{175}$$

Hence

$$\frac{d^2\Psi}{dz^2} = ik \frac{d}{dz}(u - v) + ik'(u - v). \tag{176}$$

From Equations (174), (175) and (176) we obtain

$$\frac{d^2\Psi}{dx^2} + k^2\Psi = f(z) \tag{177}$$

which is the original wave equation we started with. Let us now consider the boundary conditions

$$u(Y) = 1 \quad \text{and} \quad v(X) = 0. \tag{178}$$

We can write down the equations governing the finite order scattering solutions u_n and v_n. In the absence of the source term, we have the u beam of only even orders and v beam of odd orders corresponding to the boundary condition (178). In the present case we can easily see that we can have u and v beams of all orders. Hence,

$$\frac{du_n}{dz} = Au_n + Bv_{n-1} + \frac{f}{2ik}\delta_{n,\,0} \tag{179}$$

$$-\frac{dv_n}{dz} = Cu_{n-1} + Dv_n + \frac{f}{2ik}\delta_{n,\,0} \tag{180}$$

with A, B, C, and D having the same meaning as before, and the initial condition

$$u_n(x,\, x) = \delta_{n,\,0} \qquad v_n(x,\, x) = 0 \tag{181}$$

for all n.

It is now a simple matter to find the u_n and v_n beams successively. Routine calculations yield

$$u_0(z) = \left[\frac{k(Y)}{k(z)}\right]^{\frac{1}{2}} \exp i \int_Y^z k(s)\, ds \; +$$

$$+ \frac{1}{2[k(z)]^{\frac{1}{2}}} \int_Y^z ds\, \frac{f(s)}{i[k(s)]^{\frac{1}{2}}} \exp\left(i \int_s^z k(t')\, dt'\right) \tag{182}$$

and

$$v_0(z) = \frac{1}{2[k(z)]^{\frac{1}{2}}} \int_z^X dt'\, \frac{f(t')}{i[k(t')]^{\frac{1}{2}}} \exp\left(i \int_z^{t'} k(s)\, ds\right). \tag{183}$$

The JWKB approximation for u_0 has one additional term which is the particular solution. V_0 is non-zero contrary to the case when there is no source term. The coupled integral equations for all $n \geq 1$ have the same form as obtained in the absence of f(z) (Equations (42) and (43)).

$$u_n(z) = \frac{1}{2[k(z)]^{\frac{1}{2}}} \int_Y^z dt\, \frac{k'(t)}{[k(t)]^{\frac{1}{2}}}\, v_{n-1}(t) \exp\left(i \int_t^z k(s)\, ds\right) \tag{184}$$

$$v_n(z) = -\frac{1}{2[k(z)]^{\frac{1}{2}}} \int_z^X dt\, \frac{k'(t)}{[k(t)]^{\frac{1}{2}}}\, u_{n-1}(t) \exp\left(i \int_z^t k(s)\, ds\right). \tag{185}$$

Summing up over all the values n we obtain the coupled set of integral equations for

$$u = \sum_{n=0}^{\infty} u_n \quad \text{and} \quad v = \sum_{n=0}^{\infty} v_n \quad \text{as:}$$

$$u(z) = u_0(z) + \frac{1}{2[k(z)]^{\frac{1}{2}}} \int_Y^z dt \, \frac{k'(t)}{[k(t)]^{\frac{1}{2}}} v(t) \exp i \int_t^z k(s)\, ds \tag{186}$$

$$v(z) = v_0(z) - \frac{1}{2[k(z)]^{\frac{1}{2}}} \int_z^X dt \frac{k'(t)}{[k(t)]^{\frac{1}{2}}} u(t) \exp i \int_z^t k(s)\, ds. \tag{187}$$

Contrast this with the coupled set of integral equations for u and v in the absence of the source term (Equations (56) and (57) earlier).

We will deal with a more rigorous method of deriving this in Chapter VI and the convergence criteria for these sets of u_n and v_n can be analyzed, which will yield the proper conditions not only for k and k' but also for f(z).

REFERENCES

[1] Bremmer, H., 'The WKB Approximation as the First Term of a Geometric Optical Series', Theory of Electromagnetic Waves Symposium, Interscience Publishers, Inc., New York, 1951, 169.

[2] Bellman, R., and R. Kalaba, 'Functional Equations, Wave Propagation and Invariant Imbedding', Journal of Mathematics and Mechanics 8 (1959), 683.

[3] Stokes, G., Mathematical and Physical Papers of Sir George Stokes, Vol. 14, Cambridge, (1904), 145.

[4] Bellman, R. and G.M. Wing, Introduction to Invariant Imbedding, John Wiley & Sons, Inc., New York, 1975.

[5] Atkinson, F.V., 'Wave Propagation and the Bremmer Series', Journal of Mathematical Analysis and Applications 1 (1960), 255.

[6] Taylor, A.E., Introduction to Functional Analysis, John Wiley & Sons, Inc., New York, 1958.

[7] Bellman, R., Stability Theory of Differential Equations, McGraw Hill, New York, 1953. References in this volume may be useful.

[8] Scott, M., Invariant Imbedding and Its Applications to Ordinary Differential Equatins, Addison-Wesley Publishing Company, London, 1973.
[9] Bellman, R., R. Vadusevan, and S. Ueno, 'On the Matrix Riccati Equation of Transport Processes'. Journal of Mathematical Analysis and Applications 44 (1973), 472.
[10] Bellman, R., and R. Vasudevan, 'Wave Equations with Sources, Invariant Imbedding and Bremmer Series Solutions', Journal of Mathematical Analysis and Applications 48 (1974), 17.
[11] Wing, G.M., 'Invariant Imbedding and Generalization of JWKB Method and the Bremmer Series'. Journal of Mathematical Analysis and Applications 48 (1974), 400.
[12] Morse, P.M. and H. Feshbach, Methods of Theoretical Physics, Mc.Graw Hill, New York, 1953.

CHAPTER V

THE BREMMER SERIES

1. INTRODUCTION

In this chapter we deal with modifications of Bremmer series based on physical concepts relating to the way in which the profile of the refractive index in each layer of the medium is approximated. In the models studied in the previous chapter the refractive index or k(x) has discontinuities at each interface. A model of the strata of the medium in which the refractive index is continuous but has discontinuities of their first derivatives at each interface was studied by Sluijter [1]. This leads to a more refined series which should converge better than the Bremmer series. Of course, these generalizations can be shown to correspond to different types of splitting of the original wave into an upgoing and downgoing component as demonstrated by Wing [2].

2. A NEW TYPE OF REFRACTIVE INDEX PROFILE IN EACH LAYER AND THE REFLECTION AND TRANSMISSION COEFFICIENTS

In the previous chapter, to obtain the solutions for the wave propagating in an inhomogeneous medium, we imagined the medium to be made up of a sequence of thin layers with constant refractive index k_s in each layer, with discontinuities in k at each interface. Within each layer the solution of the wave function is exactly known.
Sluijter [1] made an improvement over this model by assuming an inhomogeneous refractive index profile in each layer whose solution is also exactly known. We know from Equation (13) of Chapter II that for a refractive index profile given by

$$k(x) = \frac{1}{(\alpha x + \beta)^2} \tag{1}$$

the JWKB solution namely

$$\Psi(x) = \frac{1}{\sqrt{k}} \exp\left(\pm i \int kdx\right) \tag{2}$$

is the exact solution. Since the profile (1) for each layer has two parameters α and β they can be chosen so that the refractive index is continuous at each interface allowing for discontinuities in the first derivatives (and higher derivatives also) as we move from layer to layer. Of course, the reflection and transmission coefficients at each interface are different from those given by Equations (18) and (21) of Chapter IV. The case of pure reflections and transmissions due to discontinuities in the first derivative has been studied [3] by Burman. Thus this type of layered model of the medium, with exact solutions known in each layer can provide a better basis than approximating the medium by layers with discontinuous refractive index.

The medium extends from x_0 to x or to ∞ as the case may be, and it is considered to be made up of a series of inhomogeneous layers $x_0 \leq z < z_1$, $z_1 \leq z < z_2 \cdots$, etc. In each layer the refractive index is defined by k(z) different for successive layers, but having the same profile as (1), so that the JWKB solutions are exact within each layer. For the layer $z_{s-1} < z < z_s$

$$k(z) = \frac{1}{(\alpha_s z + \beta_s)^2} \tag{3}$$

and similar expressions hold for $s = 0, 1, 2, \cdots$, etc. α_s and β_s can be adjusted so that k remains continuous at the interfaces $s = 0, 1, 2, \cdots$, etc. However, $k'(z)$ at the interfaces undergo discontinuities.

Just as we calculated the reflection and transmission factors at the interface of two layers with constant k values in Chapter IV (Equations (5) and (6)), exploiting the principle that the wave and its first derivative are continuous at the interface, we can obtain the R and T, the reflection and transmission factors for a JWKB type wave going from one layer to the next, keeping the continuity of the k value at the interface. k in each layer has the profile as in Equation (1). The first derivative has a jump. Following the procedure in Chapter I, the condition that Ψ and Ψ' are continuous yields the following expressions for T and R at the interface

$$T(z)_{\text{interface}} = \frac{2ik^{\frac{1}{2}}}{2ik^{\frac{1}{2}} + (\ell_2 - \ell_1)_{\text{interface}}} , \tag{4}$$

where $\ell_{1,2} = \left(\frac{dk^{-\frac{1}{2}}}{dx}\right)_{1,2}$ pertaining to the two layers 1 or 2 taken at the interface. Since k is continuous at the interface, k in (4) is the value of the refractive index at the interface. Similarly,

$$R(z)_{\text{interface}} = \frac{-(\ell_2 - \ell_1)}{2ik^{\frac{1}{2}} + (\ell_2 - \ell_1)} . \tag{5}$$

(For the profile of Equation (3) we can see that $\ell_1 = \alpha_1$ and $\ell_2 = \alpha_2$ when the wave tries to go from layer 1 to layer 2 i.e. for right moving beam.) It is evident that

$$1 + R = T \tag{6}$$

and that the conservation of current conditions holds

$$RR^* + TT^* = 1. \tag{7}$$

If we now want to write the forward differential equations for the upgoing and downgoing waves u and v as we did in Equations (50) and (51) of Chapter IV, we now have for the JWKB wave the following amplitude within each layer

$$u = \Psi_+ = \exp\left[i\int k\, ds - \int \frac{k'}{2k}\, ds\right]. \tag{8}$$

$$v = \Psi_- = \exp\left[-i\int k\, ds - \int \frac{k'}{2k}\, ds\right]. \tag{9}$$

Carrying out the input-output analysis in the region z + Δ, and z − Δ, around z, we can easily see that the coefficients a, b, c and d which have the same meaning as in Equations (45), (46), (47), (48), and (49) of Chapter IV are given by

$$a = \frac{k'}{2k} - ik\left\{1 + \frac{1}{2k^2}\left[\frac{3}{4}\frac{k'^2}{k^2} - \frac{k''}{2k}\right]\right\}. \tag{10}$$

$$b = ik\left\{\frac{1}{2k^2}\left[\frac{3}{4}\frac{k'^2}{k^2} - \frac{k''}{2k}\right]\right\}. \tag{11}$$

Similar expressions can be found for c and d, and we arrive at the differential equations satisfied by u and v as

$$\frac{du}{dz} = \left[\frac{-k'}{2k} + ik(1+\delta)\right] u + ik\delta v , \tag{12}$$

$$\frac{dv}{dz} = \left[\frac{-k'}{2k} - ik(1+\delta)\right] v - ik\delta u , \tag{13}$$

where

$$\delta = \frac{1}{2k^2}\left[\frac{3}{4}\frac{k'^2}{k^2} - \frac{k''}{2k}\right]. \tag{14}$$

If we had started with a differential equation with a large factor λ^2 multiplying the refractive index as in

$$\Psi'' + \lambda^2 k^2(z)\Psi = 0 \tag{15}$$

we could obtain the set (12) and (13) with a slight modification*

$$\frac{du}{dz} = \left[\frac{-k'}{2k} + i\lambda k(1+\bar{\delta})\right] u + i\lambda k\bar{\delta} v , \tag{16}$$

$$\frac{dv}{dz} = \left[\frac{-k'}{2k} - i\lambda k(1+\bar{\delta})\right] v - i\lambda k\bar{\delta} u . \tag{17}$$

The JWKB solution for unit input at the left end is given by

$$u_0(z) = \left[\frac{k(0)}{k(z)}\right]^{\frac{1}{2}} \exp\left[i\lambda \int_0^z k(s)(1+\bar{\delta}(s))\, ds\right]. \tag{18}$$

which is obtained by ignoring the reflected flux v completely. In the usual Bremmer series, the first term which is called the JWKB approximation, all reflections of the Fresnel type

---

* When $\lambda \neq 1$ it is to be noted that Equation (14) is to be multiplied by $1/\lambda^2$ to obtain $\bar{\delta}$.

are neglected, while the JWKB solution (18) obtained above ignores all reflections of the Burman type [3].

If we now want to obtain the v flux at x after having experienced one reflection inside the medium, we can label it as $v_1(x)$ and write down at z = x

$$v_1(x) = i\lambda k^{-\frac{1}{2}}(x) \int_x^L ds\, b(s) u_0(s) \exp\left[i\lambda \int_x^s d\sigma\, k(\sigma)(1 + \bar{\delta}(\sigma)\right]$$

$$= i\lambda k^{-\frac{1}{2}}(x) \int_x^L ds\, k^{3/2}(s)\bar{\delta}(s) u_0(s) \times \exp\left[i\lambda \int_x^s d\sigma\, k(\sigma)(1 + \bar{\delta}(\sigma)\right] \qquad 0 \le x \le L \tag{19}$$

if the medium extends from 0 to L.

In the above $u_0(s)$ gives the direct beam up to s which undergoes one reflection at s, and, hence, gets multiplied by b(s), the reflection factor and travels down to x with no more reflections. We integrate over s, since this can happen at any point above x, the upper limit L being the extension of the medium which may go to ∞ as the case may be. Of course, the boundary condition assumed is that no beam enters the medium at the right extremity L.

Similarly, the twice reflected beam moving to the right at z = x, $u_2(x)$ can be obtained if v_1 experiences one more reflection at any point s below x inside the medium. Hence

$$u_2(x) = i\lambda k^{-\frac{1}{2}}(x) \int_0^x ds\, k^{3/2}(s)\bar{\delta}(s) v_1(s) \exp\left[i\lambda \int_s^x k(\sigma)(1 + \bar{\delta}(\sigma))\, d\sigma\right]. \tag{20}$$

Hence, in general, we can obtain the following integral equations

$$u_{2N}(x) = i\lambda k^{-\frac{1}{2}}(x) \int_0^x ds\, k^{3/2}(s)\bar{\delta}(s) v_{2N-1}(s) \exp\left[i\lambda \int_s^x k(\sigma)(1 + \bar{\delta}(\sigma))\, d\sigma\right]. \tag{21)*$$

$$v_{2N+1}^{(x)} = i\lambda k^{-\frac{1}{2}}(x) \int_x^L ds\, k^{3/2}(s)\bar{\delta}(s) u_{2N}(s) \exp\left[i\lambda \int_x^s d\sigma\, k(\sigma)(1 + \bar{\delta}(\sigma))\right]. \tag{22}$$

with N = 1, 2, 3, ..., for u_{2N}.

Equivalently, we can obtain the differential equations satisfied by the u and v streams of different orders by differentiating (21) and (22) with respect to x:

$$\frac{du_{2N}}{dx} = \left\{- \frac{k'}{2k} + i\lambda k(1 + \bar{\delta})\right\}u_{2N} + i\lambda k\bar{\delta}v_{2N-1}. \tag{23}$$

$$\frac{dv_{2N+1}}{dx} = \left\{- \frac{k'}{2k} - i\lambda k(1 + \bar{\delta})\right\}v_{2N+1} - i\lambda k\bar{\delta}u_{2N}. \tag{24}$$

The total u and v flux is given by

$$u = \sum_{N=0}^{\infty} u_{2N}$$
$$v = \sum_{N=0}^{\infty} v_{2N+1} \tag{25}$$

and u and v satisfy Equations (16) and (17). We can easily see that $\Psi = u + v$ the total solution can be written in terms of u and v as

$$\Psi' + (\frac{k'}{2k} + i\lambda k)\Psi = 2i\lambda ku. \tag{26}$$

$$\Psi' + (\frac{k'}{2k} - i\lambda k)\Psi = -2i\lambda kv. \tag{27}$$

Therefore,

$$\Psi' + \frac{k'}{2k}\Psi = i\lambda k(u - v) \tag{28}$$

differentiating this with respect to z, and substituting for Ψ' from (28), we obtain

$$\Psi'' + \left(\frac{k'}{2k} - \frac{3}{4}\frac{k'^2}{k^2}\right)\Psi = i\lambda\,\frac{k'(u - v)}{2} + i\lambda k(u' - v'). \tag{29}$$

In view of Equations (16), (17), and (25), we find

$$u' - v' = \frac{-k'}{2k}(u - v) + i\lambda k(1 + 2\bar{\delta})\Psi. \tag{30}$$

Substituting (30) into (29), we find

$$\Psi'' + \left(\frac{k''}{2k} - \frac{3}{4}\frac{k'^2}{k^2}\right)\Psi = -\lambda^2 k^2 \Psi - 2\lambda^2 k^2 \bar{\delta}\Psi \tag{31}$$

which in view of the definition of $\bar{\delta}$ yields

$$\Psi'' + \lambda^2 k^2 \Psi = 0 \tag{32}$$

which is the original second order equation we started with (15).

The differential Equations (16) and (17) or the equations resulting after summing up over all N values of the simultaneous set of integral Equations (21) and (22) lead to a coupled set of Fredholm type integral equations.

For the given initial condition and assuming L goes to ∞, we have

$$u(x) = y(x) + \int_0^x ds\; y(x, s)v(s)b(s) \tag{33}$$

$$v(x) = \int_x^\infty ds\; y^*(x, s)d(s)u(s), \tag{34}$$

where

$$y(x, s) = y(x)y^{-1}(s) \tag{35}$$

and

$$y(x) = k^{-\frac{1}{2}}(x) \exp\left[i\lambda \int_0^x k(\sigma)(1 + \bar{\delta}(\sigma))\, d\sigma\right] \tag{36}$$

y^* is the complex conjugate of y and

$$b = d = i\lambda k\bar{\delta}\;. \tag{37}$$

If we assume further that

$$\int_0^\infty |k\delta|\; dx < \infty \tag{38}$$

as stated by Sluijter [1], we can follow Atkinson's arguments given in the earlier chapter and arrive at the conclusion that (33) and (34) have a unique solution in the class of functions that are continuous and bounded unless the homogeneous set (with $y(x)$ replaced by zero above in (33)) have nontrivial solutions, i.e., u and v not identically zero.

To preclude the last possibility mentioned above we consider the homogeneous set

$$u(x) = \ell \int_0^x dsy(x, s)b(s)v(s) \qquad (39)$$

$$v(x) = \ell \int_x^\infty dsy^*(x, s)d(s)u(s) \qquad (40)$$

and can proceed as Atkinson [5] did and demonstrate that the set has no real eigenvalues. Again writing the integral Equations (33) and (34) with the parameter ℓ, before the integrals, we can show that the Neumann series converges if $|\ell| < |\ell_1|$ where ℓ_1 is the smallest eigenvalue of the homogeneous set (39 and 40). Investigating the conditions to be imposed on b(s) or d(s) in order that $|\ell_1| > 1$, we are led to the following criterion, for absolute and uniform convergence [1]

$$\int_0^\infty |k(x)\bar{\delta}(x)|dx < \frac{\Pi}{2\lambda} . \qquad (41)$$

If we now transform the independent and dependent variables in (26) and (27)

$$d\xi = \lambda k \; dx \qquad (42)$$

and

$$(U_\uparrow + U_\downarrow) = U = k^{+1/2}(u + v) = k^{1/2}\psi \qquad (43)$$

we obtain

$$U_\uparrow = \frac{-i}{2}\frac{dU}{d\xi} + \frac{U}{2} \qquad (44)$$

$$U_\downarrow = \frac{i}{2}\frac{dU}{d\xi} + \frac{U}{2} . \qquad (45)$$

Also, Equations (16) and (17) can be rewritten as

$$\frac{dU_\uparrow}{d\xi} = iU_\uparrow + i\hat{\delta}(\xi)U \tag{46}$$

$$\frac{dU_\downarrow}{d\xi} = -iU_\downarrow - i\hat{\delta}(\xi)U \tag{47}$$

$\hat{\delta}$ is the function $\bar{\delta}$ expressed in the new variable ξ. Differentiating (44) and (45) once with respect to ξ and making use of expressions (46) and (47), we eliminate $dU_\uparrow/d\xi$ and $dU_\downarrow/d\xi$ and obtain

$$\frac{d^2U}{d\xi^2} + (1 + 2\hat{\delta})U = 0. \tag{48}$$

$2\hat{\delta}$ expressed in the new variable ξ, is equivalent to the expression

$$2\hat{\delta} = \left(\frac{1}{4}\frac{k'^2}{k^2} - \frac{k''}{2k}\right), \tag{49}$$

where the differentiations are with respect to ξ. We also obtain Equation (48) performing Liouville transformations (42) and (43) on the original second order Equation (32). Now one can imagine the layers in the medium to have constant values of $(1 + 2\hat{\delta})$ within each stratum, and go on to construct a Bremmer like series as described earlier and obtain another series [4]. These generalizations are a natural consequence of the different ways of splitting the wave function Ψ into up and down going streams as we see in the next section [3].

3. SPLITTING OF THE WAVE FUNCTION

The various generalizations of the Bremmer series can be obtained by imagining the total wave function Ψ as being made up of different types of u and v beams described earlier. This has been illustrated by Wing [3] and also in Reference [6]. Starting with the second order wave equation (15) λ being a large parameter, our task is to split Ψ into a pair of u and v such that

$$\Psi(z) = u + v \tag{50}$$

$$\Psi'(z) = \alpha(z)u + \beta(z)v \tag{51}$$

with α and β being differentiable. Making use of the wave equation we solve for u' and v'

$$\frac{du}{dz} = \frac{1}{(\beta - \alpha)} \{[\alpha' + \alpha\beta + \lambda^2 k^2]u + [\beta' + \beta^2 + \lambda^2 k^2]v\} \tag{52}$$

$$-\frac{dv}{dz} = \frac{1}{(\beta - \alpha)}\{[\alpha' + \alpha^2 + \lambda^2 k^2]u + [\beta' + \alpha\beta + \lambda^2 k^2]v\}. \tag{53}$$

We can have a finite extension of the medium $0 \leq z \leq x$, within which $(\beta - \alpha) \neq 0$. This condition must be imposed on the problem. We may have the boundary conditions $u(0) = 1$ and $v(x) = 0$. This produces only one solution to the Equation (15) as is easily seen.

We recognize that the set (52) and (53) correspond to the transport equations of the previous chapter (Equations (126) and (127)) if we agree to name the coefficients as A, B, C, and D given by

$$A = \frac{1}{(\beta - \alpha)} [\alpha' + \alpha\beta + \lambda^2 k^2] \tag{54}$$

$$B = \frac{1}{(\beta - \alpha)} [\beta' + \beta^2 + \lambda^2 k^2] \tag{55}$$

$$C = \frac{1}{(\beta - \alpha)} [\alpha' + \alpha^2 + \lambda^2 k^2] \tag{56}$$

$$D = \frac{1}{(\beta - \alpha)} [\beta' + \alpha\beta + \lambda^2 k^2]. \tag{57}$$

As has been illustrated in the previous section, the set of equations for u and v can easily lead to the order of scattering equations for u_{2N} and v_{2N+1}

$$\frac{du_{2N}}{dz} = Au_{2N} + Bv_{2N-1} \tag{58}$$

$$-\frac{dv_{2N+1}}{dz} = Cu_{2N} + Dv_{2N+1} \tag{59}$$

with $u(0) = \delta_{N,0}$ and $v_{2N+1}(x) = 0$ for all N. u_{2N} and v_{2N+1} have the physical meaning that they represent the stream that has undergone exactly the number of backscatterings indicated by their indices. $(1 - \bar{A}\Delta)^*$ and $(1 - \bar{D}\Delta)^*$ are the fractions of u and v beams that are transmitted in their direction of flow forward for the u and v beams, while $B\Delta$ and $c\Delta$ are the reflected ratios in traversing a path Δ at z.

To obtain the classical Bremmer case, we assume the splitting corresponding to

$$\alpha(z) = i\lambda k(z) \quad \text{and} \quad \beta(z) = -i\lambda k(z) \tag{60}$$

provided $k(z) > 0$ is differentiable in the entire region. We have the transport equations for u and v given by (making use of Equations (54) through (57)),

$$\frac{du}{dz} = (i\lambda k - \frac{k'}{2k})u + \frac{k'}{2k}v \tag{61}$$

$$-\frac{dv}{dz} = -\frac{k'}{2k}u + (i\lambda k + \frac{k'}{2k})v. \tag{62}$$

The order of scattering equations correspond to those given by (58) and (59). Neglecting the non-diagonal parts, we obtain the usual JWKB solution for u and v. For the given boundary conditions we obtain the usual coupled integral equations for u and v. Also making use of the differential equations for u_{2N} and v_{2N+1} we can show that the classical Bremmer series is obtained as was done in Chapter IV. The first term of that series is the JWKB approximation of Equation (37) of Chapter IV.

4. EXTENSIONS TO OTHER TYPES OF SERIES

From the structure of the iterative Equations (58) and (59) for u_{2N} and v_{2N+1}, it is desirable to have B and C as small as possible to get better approximations with a smaller number of succeeding terms for the u and v series.

If we attempt to realize such choices of α and β that B and C are each equal to 0, we require:

$$\alpha'(z) + \alpha^2(z) + \lambda^2 k^2 = 0 \tag{63}$$

$$\beta'(z) + \beta^2(z) + \lambda^2 k^2 = 0. \tag{64}$$

We try to solve these equations approximately by means of a formal series for α and β. We write for α

---

* $\bar{A} = -A$ and $\bar{D} = -D$ with A and D occurring in Equations (58) and (59).

$$\alpha = \lambda\left[a_0(z) + \frac{a_1}{\lambda} + \frac{a_2}{\lambda^2} + \cdots\right]. \tag{65}$$

Substituting this into Equation (63) and equating like powers of λ, we obtain

$$a_0 = \pm\, ik, \tag{66}$$

$$a_0' = -a_1(2a_0), \tag{67}$$

$$a_1' = -\,[a_1^2 + 2a_0a_2], \tag{68}$$

$$a_2' = -\,2[a_1a_2 + a_0a_3]. \tag{69}$$

In Chapter II we wrote down the Eikonal expansion, for the solution of the Schrödinger equation identical with Equation (65) and obtained the Riccati equation satisfied by the integrand of the phase, namely, $(\lambda S')$. (Equation 3 of Chapter II.) This quantity obeys the same Riccati equation as (63) or (64) and it was also expanded as is done in Equation (65). Hence, we have

$$a_0 = \pm\, ik \tag{70}$$

$$a_1 = -\frac{k'}{2k} \tag{71}$$

$$a_2 = \mp\frac{1}{2ik}\left(\frac{3}{4}\frac{k'^2}{k^2} - \frac{k''}{2k}\right) \tag{72}$$

$$a_3 = -\frac{1}{8}\left(\frac{k'''}{k^3} - \frac{6k'k''}{k^4} + \frac{6k'^3}{k^5}\right) \tag{73}$$

and so on. A similar expansion for β as a solution of Equation (66) can be formulated

$$\beta = \lambda\left[b_0 + \frac{b_1}{\lambda} + \frac{b_2}{\lambda^2} + \cdots\right] \tag{74}$$

and the same expressions as in Equations (70) to (73) will be obtained for the successive terms $b_0, b_1, b_2, \cdots$,. However, if the condition $\beta - \alpha \neq 0$ holds, we have $a_0 = +\, i\lambda k$ and

$b_0 = -i\lambda k$, so that $\beta - \alpha = -2i\lambda k$ which is never zero so long as k remains greater than zero. We have seen already that if we stop with the first term a_0 and b_0 for approximating α and β respectively we obtain Equation (61) and (62) which are the classical Bremmer equations.

If we take the next approximation and $\alpha = \lambda a_0 + a_1$ and $\beta = \lambda b_0 + b_1$, and equate to zero all the higher terms, we easily see from the values of a_1 and b_1 from Equations (70) and (71) that we obtain

$$\alpha(z) = i\lambda k - \frac{k'}{2k} \quad \text{and} \quad \beta(z) = -i\lambda k - \frac{k'}{2k}\,. \tag{75}$$

Therefore, we obtain the following expressions for the coefficients A, B, C, and D using the expression (54) through (57)*.

$$A = i\lambda k - \frac{k'}{2k} - \frac{1}{2i\lambda k}\left[\frac{3}{4}\frac{k'^2}{k^2} - \frac{k''}{2k}\right] = -\frac{k'}{2k} + i\lambda k(1 + \bar{\delta}) \tag{76}$$

$$D = -A^* \tag{77}$$

$$B = -\frac{1}{2i\lambda k}\left(\frac{3}{4}\frac{k'^2}{k^2} - \frac{k''}{2k}\right) = i\lambda k\bar{\delta} \tag{78}$$

$$C = B \tag{79}$$

$$\bar{\delta} = \frac{1}{2k^2\lambda^2}\left(\frac{3}{4}\frac{k'^2}{k^2} - \frac{k''}{2k}\right) \tag{80}$$

Hence, the transport equations become

$$\frac{du}{dz} = \left[-\frac{k'}{2k} + i\lambda k(1 + \bar{\delta})\right]u + i\lambda k\bar{\delta}v \tag{81}$$

$$-\frac{dv}{dz} = \left[\frac{k'}{2k} + i\lambda k(1 + \bar{\delta})\right]v + i\lambda k\bar{\delta}u. \tag{82}$$

Equations (16) and (17) are identical with the above set (81) and (82). Sluijter [2] used the physical idea of taking the elementary layers having a profile such that the JWKB solutions are exact in each layer and obtained the set (16) and (17) as demonstrated earlier. Thus we see that this refinement

---

* It is to be noted that capital letters C and D are not the same as small letters c and d used earlier in Chapter IV.

of Sluijter [1] is equivalent to incorporating one more term in the coefficients α and β which characterize the splitting of the original wave. The terms in the series for α and β also correspond to the terms of the Eikonal expansion of the two components of the wave.

The JWKB approximation for the solution is now

$$u(x) = \left(\frac{k(0)^{\frac{1}{2}}}{k^{\frac{1}{2}}(x)}\right) \exp\left[i\lambda \int_0^x k(s)(1 + \bar{\delta}(s))ds\right] \tag{83}$$

for the given initial condition. However, one can easily see that the coefficient A of the transport equation (81) is obtained from the first three terms of the Eikonal expansion.

$$A = i\lambda k - \frac{k'}{2k} + i\lambda k\bar{\delta} \quad \text{with } \alpha = \lambda\left(a_0 + \frac{a_1}{\lambda} + \frac{a_2}{\lambda^2}\right). \tag{84}$$

Where a_0, a_1, and a_2 are given by Equations (70), (71), and (72), respectively. In the next approximation for the splitting of the wave we are to include for $\alpha(z)$ these three terms which constitute the present coefficient A in the transport Equations. That is, if we go to the next higher approximation we can have the splitting:

$$\Psi = \bar{u} + \bar{v} \tag{85}$$

$$\Psi' = \bar{\alpha}\bar{u} + \bar{\beta}\bar{v}, \tag{86}$$

where $\bar{\alpha}$ and $\bar{\beta}$ are to be taken as

$$\bar{\alpha} = A \tag{87}$$

of Equation (84) and

$$\bar{\beta} = A^* = -D \tag{88}$$

of Equation (80). Also, we should notice that C given by the expression (56) is related to A in a simple fashion

$$C = \frac{\alpha' + \alpha^2 + \lambda^2 k^2}{(\beta - \alpha)} = \frac{\alpha' + \alpha\beta + \lambda^2 k^2 + \alpha^2 - \alpha\beta}{(\beta - \alpha)} = A - \alpha. \tag{89}$$

Similarly,

$$B = D + \beta. \tag{90}$$

Thus we can go on to include in the coefficients α and β higher order terms in their series expansion and obtain different splittings. For each type of splitting the transport coefficients A and D can be obtained by the use of Equations (54) through (57). The reflection functions C and B are easy to evaluate for each splitting.

The method of obtaining the successive terms of the Eikonal expansion (or expansion in the phase) is given in Equation (6) of Chapter II. It is therefore easy to refine the JWKB solutions up to any high order of accuracy by splitting the wave suitably as described earlier. Each type of splitting leads to a set of transport equations. By using the now familiar argument, the transport equation for u_{2N} and v_{2N+1} of any order of scattering solutions can be written down. Correspondingly, we can obtain the coupled set of integral equations related to each manner of splitting and hence obtain relevant generalized Bremmer series without any difficulty.

REFERENCES

[1] Sluijter, F.W., 'Generalizations of the Bremmer Series Based on Physical Concepts', Journal of Mathematical Analysis and Applications 27 (1969), 282.
[2] Wing, G. Milton, 'Invariant Imbedding and Generalizations of the WKB Method and Bremmer Series', Journal of Mathematical Analysis and Applications 48 (1974), 400.
[3] Burman, R., Journal Math. and Phys. 45 (1966), 162. Proc. IEEE 54 (1966), 887.
[4] Polischuk, I.M., 'Radio Engineering Physics', U.S.A., 11 (1966), 1975.
[5] Atkinson, F.V., Journal of Mathematical Analysis and Applications 1 (1960) 255.
[6] Bellman, R. and R. Vasudevan, 'On the Generalization of Bremmer Series Solutions of Wave Equations', Journal of Mathematical Analysis and Applications 52 (1975), 151.

CHAPTER VI

GENERALIZATIONS

1. INTRODUCTION

In this chapter we describe the method of obtaining the various generalizations of the Bremmer series by a successive diagonalization procedure of the transport equations for the u and v beams obtained at each stage [1]. This was carried out by various authors [2], [3], and [4] for the wave equation, without the source term. We study these methods and see how a unified picture of these procedures emerges. We also illustrate how these are related to the method of applying the Liouville transformations successively to the wave equation [5]. We also treat the wave equation with a source term and demonstrate the technique of arriving at different types of generalized Bremmer series. The diagonalizing matrix at each successive stage of the process represents the manner in which the wave function can be split up at that stage. As we have seen already in Chapter V, better and better approximations can be obtained for the wave function even with the first term of the generalized Bremmer series, if the splitting of the wave function involves higher and higher terms of the Eikonal approximation [6]. It is seen that for each term to be finite, it is imperative that the $k(x)$ should be nonzero, i.e., the refractive index should remain positive in the region of interest. Otherwise, we meet with what is known as the turning point difficulty. It has also been shown earlier that the successive terms of the Eikonal approximation results by solving the Riccati equation in an iterative fashion. However, if we solve the Riccati equation by quasi-linearization [7], even the first approximation can be shown to include an infinite sum of certain class of terms of the Eikonal series. This avoids the turning point difficulty as we shall see below.

2. METHOD OF SUCCESSIVE DIAGONALIZATION

Let us start with the wave equation

$$\frac{d^2\Psi}{dx^2} + \lambda^2k^2(x)\Psi = 0 \tag{1}$$

and write it down in a vector matrix form

$$\frac{d}{dx}\begin{pmatrix}\Psi\\ \Psi'\end{pmatrix} = \frac{d}{dx}\vec{\Psi} = \begin{pmatrix}0 & 1\\ -\lambda^2k^2, & 0\end{pmatrix}\begin{pmatrix}\Psi\\ \Psi'\end{pmatrix} = A\ \vec{\Psi}. \tag{2}$$

The matrix that diagonalizes A is the matrix P_0 given for any particular value of x.

$$P_0 = \begin{bmatrix}1 & 1\\ i\lambda k & -i\lambda k\end{bmatrix} \tag{3}$$

such that

$$P_0^{-1}AP_0 = \begin{bmatrix}i\lambda k & 0\\ 0 & -i\lambda k\end{bmatrix}. \tag{4}$$

We have assumed that $k(x) > 0$ for all x in the interval in which we are interested.

We now make a transformation of the vector $\vec{\Psi}$ using $\underline{P}_0$.

$$\vec{\Psi} = P_0\ \vec{u}_0 \quad \text{with } \vec{u}_0 = \begin{pmatrix}u_0\\ v_0\end{pmatrix} \tag{5}$$

Of course the above equation is equivalent to splitting the wave function into components u_0 and v_0 such that

$$\Psi = u_0 + v_0 \tag{6}$$

$$\Psi' = \alpha(x)u_0 + \beta(x)v_0, \tag{7}$$

where $\alpha(x) = i\lambda k$ and $\beta(x) = -i\lambda k$ as explained in Equations (60) and (66) of Chapter V. To obtain the transport euqation for the vector $\vec{u}_0$ given by

$$\vec{u}_0 = \begin{pmatrix}u_0\\ v_0\end{pmatrix} \tag{8}$$

we carry out the following operatons on Equation (2)

$$P_0^{-1}\frac{d}{dx}\vec{\Psi} + \frac{d}{dx}[P_0^{-1}]\vec{\Psi} = P_0^{-1}AP_0P_0^{-1}\vec{\Psi} + \frac{d}{dx}[P_0^{-1}]P_0P_0^{-1}\vec{\Psi} \tag{9}$$

$$\frac{d}{dx}\begin{pmatrix}u_0\\ v_0\end{pmatrix} = \begin{pmatrix}i\lambda k & 0\\ 0 & -i\lambda k\end{pmatrix}\begin{pmatrix}u_0\\ v_0\end{pmatrix} - \frac{k'}{2k}\begin{pmatrix}1 & -1\\ -1 & 1\end{pmatrix}\begin{pmatrix}u_0\\ v_0\end{pmatrix}. \tag{10}$$

Hence, we have

$$\frac{du_0}{dx} = \left(i\lambda k - \frac{k'}{2k}\right)u_0 + \frac{k'}{2k}v_0 \tag{11}$$

$$\frac{dv_0}{dx} = -\left(i\lambda k + \frac{k'}{2k}\right)v_0 + \frac{k'}{2k}u_0\,. \tag{12}$$

The coupled set (11) and (12) are the same as those obtained for the upgoing and downgoing beams in Chapter V, Equations (61) and (62). We can express this set in a neat form as Van Kampen does by using the Pauli matrices [2].

$$\frac{d}{dx}\begin{pmatrix}u_0\\ v_0\end{pmatrix} = \left[i\lambda k\sigma_z + \frac{k'}{2k}\sigma_x - \frac{k'}{2k}I\right]\begin{pmatrix}u_0\\ v_0\end{pmatrix}. \tag{13}$$

The σ matrices are

$$\sigma_x = \begin{pmatrix}0 & 1\\ 1 & 0\end{pmatrix}; \qquad \sigma_y = \begin{pmatrix}0 & -i\\ i & 0\end{pmatrix}; \qquad \sigma_z = \begin{pmatrix}1 & 0\\ 0 & -1\end{pmatrix}.$$

If we now put

$$\begin{bmatrix}u_0\\ v_0\end{bmatrix} = \frac{1}{\sqrt{k}}\begin{bmatrix}u_1\\ v_1\end{bmatrix} \tag{14}$$

we obtain the new entities u_1 and v_1 which satisfy

$$\frac{d}{dx}\begin{pmatrix}u_1\\ v_1\end{pmatrix} = \left(i\lambda k\sigma_z + \frac{k'}{2k}\sigma_x\right)\begin{pmatrix}u_1\\ v_1\end{pmatrix} . \tag{15}$$

This is equivalent to the transformation of Reference [3], given by

$$\begin{bmatrix}\Psi\\ \Psi'\end{bmatrix} = \begin{bmatrix}k^{-\frac{1}{2}} & k^{-\frac{1}{2}}\\ i\,k^{\frac{1}{2}} & -i\,k^{\frac{1}{2}}\end{bmatrix}\begin{bmatrix}u_1\\ v_1\end{bmatrix} \tag{16}$$

Now if we choose the independent variable as x_1 by the relation

$$dx_1 = k dx \tag{17}$$

we have

$$\frac{d}{dx_1}\begin{pmatrix}u_1\\ v_1\end{pmatrix} = \left(i\lambda\sigma_z + \frac{k'}{2k}\sigma_x\right)\begin{pmatrix}u_1\\ v_1\end{pmatrix} = A_1\vec{u}_1 \tag{18}$$

where the prime denotes differentiation with respect to x_1. Hence, by differentiating once again, we obtain from Equation (18)

$$\frac{d^2\vec{u}_1}{dx^2} = A_1'\,\vec{u}_1 + A_1^2\,\vec{u}_1 \tag{19}$$

and since

$$A_1' = \left(\frac{k''}{2k} - \frac{k'^2}{2k^2}\right)\sigma_x \tag{20}$$

and

$$A_1^2 = \left(-\lambda^2 + \frac{k'^2}{4k^2}\right) I, \tag{21}$$

we can add up the two equations of the coupled set (19) and obtain

$$\frac{d^2(u_1 + v_1)}{dx_1^2} = \frac{d^2\varphi}{dx_1^2} = \left(-\lambda^2 - \frac{k'^2}{4k^2} + \frac{k''}{2k}\right)\varphi \tag{22}$$

that is

$$\frac{d^2\varphi}{dx_1^2} + \lambda^2 \left\{1 + \frac{1}{\lambda^2}\left[\frac{k'^2}{4k^2} - \frac{k''}{2k}\right]\right\}\varphi = 0 \tag{23}$$

which in turn can be written as

$$\frac{d^2\varphi}{dx_1^2} + \lambda^2 \left\{1 - \frac{1}{\lambda^2}\left[\frac{1}{4}\left(\frac{d\gamma_0}{dx_1}\right)^2 + \frac{1}{2}\frac{d^2\gamma_0}{dx_1^2}\right]\right\}\varphi = 0, \tag{24}$$

where $\gamma_0 = \log k$. Equations (23) and (24) are the transformed equations obtained from the original wave equation by the application of the Liouville transformation as demonstrated in Chapter II, Equations (23), (24) and (28).

The Bremmer series at this stage will be obtained by solving for the two components $u_{1,\,2n}$ and $v_{1,\,2n+1}$ satisfying the set of coupled differential equations represented by (18)

$$\frac{du_{1,\,2n}}{dx_1} = i\lambda u_{1,\,2n} + \frac{k'}{2k} v_{1,\,2n-1} \tag{25}$$

$$\frac{dv_{1,\,2n+1}}{dx_1} = -\,i\lambda v_{1,\,2n+1} + \frac{k'}{2k} u_{1,\,2n}. \tag{26}$$

$$Y \le x_1 \le X$$

The boundary conditions are

$$u_{1,\,2n}(Y) = \delta_{n,\,0} \quad \text{and} \quad v_{1,\,2n+1}(X) = 0 \quad \text{for all } n, \tag{27}$$

Y being the left end and X the right end of the medium. The coupled set of integral equations for any order n of the components $u_{1,\,2n}$ and $v_{1,\,2n+1}$ can be obtained as described in the earlier chapters. Summing up over all the n values we have the coupled integral equations for u_1 and v_1.

Ignoring the non-diagonal components, we obtain the WKB solutions for u_1 and v_1.

$$u_1 \sim \exp\left(i\lambda \int dx_1\right) \sim \exp\left(i\lambda \int k dx\right) \tag{28}$$

and $u_0 \sim \frac{1}{\sqrt{k}} \exp\left(i\lambda \int k dx\right)$ is the usual JWKB solution for Ψ as obtained earlier (see the preceding chapters).

We can now show that we can handle the wave equation with the source term f(x) on the right-hand side without recourse to the heuristic arguments we employed in Chapter IV to obtain Equations (172) and (173) (see Reference [8]). We start with the equation

$$\Psi'' + \lambda^2 k^2(x)\Psi = f(x) \tag{29}$$

and write down the corresponding vector-matrix equation

$$\frac{d}{dx}\begin{bmatrix}\Psi \\ \Psi'\end{bmatrix} = \begin{pmatrix}0 & 1 \\ -\lambda^2 k^2 & 0\end{pmatrix}\begin{pmatrix}\Psi \\ \Psi'\end{pmatrix} + \begin{pmatrix}0 \\ f(x)\end{pmatrix} . \tag{30}$$

Applying the diagonalizing transformation P_0^{-1} to both sides of the equation as we did to obtain Equations (9) and (10), we led to Equations (172) and (173) of Chapter IV in a routine fashion since

$$P_0^{-1}(x) = -\frac{1}{2ik}\begin{bmatrix}-i\lambda k & -1 \\ -i\lambda k & 1\end{bmatrix} . \tag{31}$$

We have

$$\frac{d}{dx}\begin{pmatrix}u_0 \\ v_0\end{pmatrix} = \left(i\lambda k\, \sigma_z + \frac{k'}{2k}\sigma_x + \frac{k'}{2k} I\right)\begin{pmatrix}u_0 \\ v_0\end{pmatrix} + \begin{pmatrix}\frac{f(x)}{2ik} \\ \frac{-f(x)}{2ik}\end{pmatrix} . \tag{32}$$

Now introducing the transformation of the independent variable and the dependent variables

$$dx_1 = k dx \quad \text{and} \quad \vec{u}_0 = k^{-\frac{1}{2}}\, \vec{u}_1 \tag{33}$$

we have

$$\frac{d}{dx_1}\begin{pmatrix}u_1 \\ v_1\end{pmatrix} = \left(i\lambda\sigma_z + \frac{k'}{2k}\sigma_x\right)\begin{pmatrix}u_1 \\ v_1\end{pmatrix} + \frac{1}{2i\lambda k^{3/2}}\begin{bmatrix}f \\ -f\end{bmatrix} \tag{34}$$

$$\frac{d}{dx_1}\vec{u}_1 = A_1\vec{u}_1 + \vec{B} . \tag{34a}$$

Differentiating again and adding up u_1'' and v_1'' we have

$$\frac{d^2\varphi_1}{dx_1^2} + \lambda^2\left\{1 - \frac{1}{\lambda^2}\left[\frac{1}{4}\left(\frac{d\gamma_0}{dx_1}\right)^2 + \frac{1}{2}\left(\frac{d^2\gamma_0}{dx_1^2}\right)\right]\right\}\varphi_1 = \frac{\bar{f}(x_1)}{k^{3/2}}. \tag{35}$$

This is the equation obtained by the Liouville transformation for the wave equation with a source term. The Bremmer series at this stage will be generated by the coupled set of differential equations for the finite order functions $u_{1,\,n}$ and $v_{1,\,n}$ taking all the values from 0 to infinity.
As we have already seen in Chapter IV, $u_{1,\,n}$ and $v_{1,\,n}$ can both be of even and odd orders, whatever the boundary conditions are, due to the presence of the source term which generates waves in either direction.
Hence,

$$\frac{du_{1,\,n}}{dx_1} = i\lambda u_{1,\,n} + \frac{k'}{2k}v_{1,\,n-1} + \frac{1}{2i\lambda k^{3/2}}\bar{f}(x_1)\delta_{n,\,0} \tag{36}$$

$$\frac{dv_{1,\,n}}{dx_1} = \frac{k'}{2k}u_{1,\,n-1} + i\lambda v_{1,\,n} - \frac{1}{2i\lambda^{3/2}}\bar{f}(x_1)\delta_{n,\,0}, \tag{37}$$

$\bar{f}(x_1)$ is the function $f(x)$ expressed in terms of the new variable x_1 and $\delta_{n,\,0}$ is the kronecker symbol. From this set (36) and (37) we can arrive at the coupled set of integral equations for $u_{1,\,n}$ and $v_{1,\,n}$ of any order and also for the total functions u_1 and v_1. The reader should refer to similar equations obtained in Chapter IV for the wave equation with the source term (see Reference [8]).

To go to the next step we rewrite Equation (35) as

$$\varphi_1'' + \lambda^2\omega_1^2\varphi_1 = f_1(x), \tag{38}$$

where

$$\lambda^2\omega_1^2 = \left(\lambda^2 + \frac{k'^2}{4k^2} - \frac{k''}{2k}\right) \tag{39}$$

and

$$f_1(x) = \frac{\bar{f}(x_1)}{k^{3/2}} . \tag{40}$$

We again perform the Liouville transformations $\varphi_2 = \omega_1^{\frac{1}{2}}\varphi_1$ and $dx_2 = \omega_1 dx_1$ and obtain by repeating the operations gone through above, the following

$$\varphi_2'' + \lambda^2\omega_2^2\varphi_2 = f_2(x_2) \tag{41}$$

with

$$\lambda^2\omega_2^2 = \left(\lambda^2 + \frac{1}{4}\frac{\omega_1'^2}{\omega_1^2} - \frac{1}{2}\frac{\omega_1''}{\omega_1}\right) \text{ and } f_2 = \frac{f_1}{\omega_1^{3/2}} \tag{42}$$

If there had been no source term $f_2(x_2) = 0$, and the WKB solution would be of the form

$$\varphi_2 \sim \exp\left(i\lambda \int dx_2\right). \tag{43}$$

Thus at each stage n of the repeated transformation process, the WKB solution will be of the form

$$\varphi_n \sim \exp\left(i\lambda \int dx_n\right). \tag{44}$$

As an alternate way of obtaining successively better and better WKB approximations, we describe an elegant method due to Van Kampen [2] which can be directly related to the successive application of the Liouville transformation with slight modifications [1].

Let us take the equation without the source term and write the transport equations for the two components u_1 and v_1 [after the first Liouville transformation] as given by Equation (18)

$$\frac{d}{dx_1}\begin{pmatrix}u_1\\v_1\end{pmatrix} = \lambda(i\sigma_z + \sigma_x \tanh\gamma_1)\begin{pmatrix}u_1\\v_1\end{pmatrix} \tag{45}$$

where $\tanh\gamma_1 = k'/2k\lambda$. Equivalently, we can write (45) as

$$\frac{d}{d\tau_1}\begin{pmatrix}u_1\\v_1\end{pmatrix} = \lambda[i\sigma_z \cosh\gamma_1 + \sigma_x \sinh\gamma_1]\begin{pmatrix}u_1\\v_1\end{pmatrix} \tag{46}$$

$$\frac{d}{d\tau_1}\vec{u}_1 = \lambda A\,\vec{u}_1; \qquad A = [i\sigma_2 \cosh\gamma_1 + \sigma_x \sinh\gamma_1]. \tag{47}$$

where $d\tau_1 = \text{sech}\,\gamma_1\, dx_1$. We now introduce a phase difference between the components by the following procedure.

$$\vec{u}_1 = \begin{bmatrix}u_1\\v_1\end{bmatrix} = e^{-\frac{i\pi}{4}\sigma_z}\,\vec{\hat{u}}_1 = \frac{1 - i\sigma_z}{\sqrt{2}} = H\,\vec{\hat{u}}_1 \tag{48}$$

$$H^{-1} = \frac{1 + i\sigma_z}{\sqrt{2}}\ . \tag{49}$$

Hence

$$\frac{d\vec{\hat{u}}_1}{d\tau_1} = \lambda H^{-1} A\, H\,\vec{\hat{u}}_1 \tag{50}$$

$$= \lambda(i\sigma_z \cosh\gamma_1 - \sigma_y \sinh\gamma_1)\vec{\hat{u}}$$

$$\frac{d\vec{\hat{u}}_1}{d\tau_1} = B\lambda\vec{\hat{u}}_1\ . \tag{51}$$

Differentiating this again with respect to τ_1 we obtain

$$\frac{d^2\vec{\hat{u}}_1}{d\tau_1^2} = \lambda^2 B^2\,\vec{\hat{u}}_1 + \lambda\frac{d\gamma_1}{d\tau_1}(i\sigma_z \sinh\gamma_1 - \sigma_y \cosh\gamma_1)\vec{\hat{u}}_1. \tag{52}$$

Adding up the two components of the coupled set of Equation (51), we easily see that (52) is equivalent to a second order equation for $\phi = \hat{u}_1 + \hat{v}_1$ given by

$$\frac{d^2\phi}{d\tau_1^2} + \frac{d\phi}{d\tau_1}\frac{d\gamma_1}{d\tau_1} + \lambda^2\phi = 0. \tag{53}$$

If we now make the transformation

$$\phi = \varphi_1 \, e^{-\gamma_1/2} \tag{54}$$

we obtain the second order equation in a familiar form.

$$\varphi_1'' + \left[\lambda^2 - \frac{1}{4}\left(\frac{d\gamma_1}{d\tau_1}\right)^2 - \left(\frac{1}{2}\,\frac{d^2\gamma_1}{d\tau_1^2}\right)\right]\varphi_1 = 0 \tag{55}$$

If we compare this with Equation (24), it is at once realized that this is equivalent to the coupled set of first order equations with suitable splitting into $\bar{u}_1$ and $\bar{v}_1$ such that

$$\frac{d}{d\tau_1}\begin{pmatrix}\bar{u}_1\\ \bar{v}_1\end{pmatrix} = \left(i\lambda\sigma_z + \frac{1}{2}\,\frac{d\gamma_1}{d\tau_1}\,\sigma_x\right)\begin{pmatrix}\bar{u}_1\\ \bar{v}_1\end{pmatrix} \tag{56}$$

In fact we derived Equation (24) from Equation (18) which is of the form (56). We will find below what these, $\bar{u}_1$ and $\bar{v}_1$, are like. Of course we can derive the set of coupled equations from (53) directly by multiplying (53) by e^{γ_1} throughout and rewrite (53) as

$$\frac{d}{d\tau_1}\left(e^{\gamma_1}\,\frac{d\phi}{d\tau_1}\right) + \lambda^2\phi e^{\gamma_1} = 0. \tag{57}$$

It can be written down in the vector matrix form as

$$\frac{d\vec{\phi}}{d\tau_1} = A\,\vec{\phi}\,, \tag{58}$$

where

$$\vec{\phi} = \begin{bmatrix}\phi\\ e^{\gamma_1}\,\dfrac{d\phi}{d\tau_1}\end{bmatrix} \quad \text{and} \quad A = \begin{bmatrix}0 & e^{-\gamma_1}\\ -\lambda^2 e^{\gamma_1} & 0\end{bmatrix}. \tag{59}$$

The diagonalizing matrix P for this A as given by Keller and Keller [3], which is equivalent to making the Liouville transformation after proper splitting, is given by

$$P = \begin{bmatrix} \alpha^{-1} & \alpha^{-1} \\ i\alpha & -i\alpha \end{bmatrix} , \tag{60}$$

where $\alpha = \lambda^{\frac{1}{2}} e^{\frac{\gamma_1}{2}}$. The eigenvalues of A are $\pm\, i\lambda$ always and we obtain

$$\frac{d\vec{\phi}_1}{d\tau_1} = \left(i\lambda\sigma_z + \frac{1}{2}\frac{d\gamma_1}{d\tau_1}\sigma_x\right)\vec{\phi}_1 \tag{61}$$

with

$$\vec{\phi} = P\,\vec{\phi}_1 . \tag{62}$$

The introduction of the phase difference was necessitated by our desire to obtain same form for the coupled set of equations for the components at each stage. Then multiplying Equation (53) by e^{γ_1} and subsequent 'flattening' transformations (60) landed us in Equation (61).

Van Kampen arrives at this Equation (61) from Equation (46) by the application of the transformation matrix

$$e^{i\pi\sigma_z/4}\, e^{\gamma_1\sigma_y/2}$$

to the $\vec{u}_1$ of Equation (46) to obtain the new vector $\vec{\phi}_1$ $e^{\gamma_1\sigma_v/2}$ rotates the A matrix of (47) into diagonalization and $e^{+\Pi/4 i\sigma_z}$ introduces the phase difference between the components.

We can see that

$$e^{\gamma_1\sigma_y/2} A e^{-\gamma_1\sigma_y/2} = i\sigma_2$$

and therefore

$$\vec{\phi}_1 = e^{i\pi\sigma_z/4}\begin{pmatrix} u_1 \\ v_1 \end{pmatrix}$$

satisfies the equation

$$\frac{d\phi_1}{d\tau_1} = \left(i\lambda\sigma_z + \frac{1}{2}\frac{d\gamma_1}{d\tau_1}\sigma_x\right)\vec{\phi}_1 .$$

The whole procedure can be carried out with a wave equation including the source term f(x) [see Reference (1)] with a little more labor.

Starting with Equation (61), we can call

$$\frac{1}{2\lambda}\frac{d\gamma_1}{d\tau_1} = \tanh \gamma_2 \tag{63}$$

and go to the next stage. Hence, by repeating the process described earlier we arrive at the equation for any n-th step as

$$\frac{d\vec{\phi}_n}{d\tau_n} = \lambda\left(i\sigma_z + \frac{1}{2\lambda}\frac{d\gamma_n}{d\tau_n}\sigma_x\right)\vec{\phi}_n \tag{64}$$

$$= \lambda(i\sigma_z + \sigma_x \tanh\gamma_{n+1})\vec{\phi}_n. \tag{65}$$

If we call

$$d\tau_{n+1} = \operatorname{sech}\gamma_{n+1} d\tau_n, \tag{66}$$

Equation (65) becomes

$$\frac{d\vec{\phi}_n}{d\tau_{n+1}} = \lambda(i\sigma_z \cosh\gamma_{n+1} + \sigma_x \sinh\gamma_{n+1})\vec{\phi}_n. \tag{67}$$

The above is of the same form as (46) and after another transformation described above we obtain

$$\frac{d\vec{\phi}_{n+1}}{d\tau_{n+1}} = \lambda\left(i\sigma_z + \frac{1}{2\lambda}\frac{d\gamma_{n+1}}{d\tau_{n+1}}\sigma_x\right)\vec{\phi}_{n+1}. \tag{68}$$

Ignoring the non-diagonal terms, we obtain at each stage the general WKB solutions of the form

$$u_{1,\, n+1} \sim \exp \pm i\lambda\int d\tau_{n+1} \tag{69}$$

and

$$\tau_{n+1} = \int_0^{\tau_n} \sqrt{1 - \frac{1}{4\lambda^2}\left(\frac{d\gamma_n}{d\tau_n}\right)^2}\, d\tau_n. \tag{70}$$

Hence we have

$$\tau_1 = \int_0^x \sqrt{k^2 - \frac{k'^2}{4\lambda^2 k^2}}\, dx; \quad \tanh \gamma_1 = \frac{k'}{2\lambda k^2} \tag{71}^*$$

$$\tanh \tau_1 = k'/2\lambda k^2 \tag{72}$$

and so on.

---

* The prime is differentiation with respect to x unlike in Equation (18).

At each stage the non-diagonal elements we neglect are of order $1/\lambda$ compared to the diagonal terms. At each step of the process the Bremmer type series can be found by writing down the corresponding equations for finite order functions for the components of the vector $\vec{\phi}_n$. These can be easily visualized by the familiar arguments given in Chapter IV and V. For example, for Equation (61) and for the vector $\vec{\phi}_1$ we have

$$\frac{d\phi_{1,\,2N}}{d\tau_1} = i\lambda\phi_{1,\,2N} + \frac{1}{2\lambda}\frac{d\gamma_1}{d\tau_1}\phi_{2,\,2N-1} \tag{73}$$

$$\frac{d\phi_{2,\,2N+1}}{d\tau_1} = \frac{1}{2}\frac{d\gamma_1}{d\tau_1}\phi_{1,\,2N} - i\lambda\phi_{2,\,2N+1} \tag{74}$$

suitable boundary conditions can be imposed and $\phi_{1,\,2N}$ and $\phi_{2,\,2N+1}$ can be solved in an iterative fashion.*

The recursive relations expressing γ_m and τ_m and

$$\tanh\gamma_{m+1} = \frac{1}{2\lambda}\frac{d\gamma_m}{d\tau_m} \; ; \quad \tau_{m+1} = \int_0^{\tau_m}\sqrt{1 - \frac{1}{4\lambda^2}\left(\frac{d\gamma_m}{d\tau_m}\right)^2}\, d\tau_m \tag{75}$$

3. APPROXIMATION TO THE EIKONAL SOLUTION USING QUASILINEARIZATION

In Chapter V (Section 2) we arrived at different kinds of splitting of the wave function Ψ by substituting for α and β a series of terms up to different orders in $(1/\lambda)$. The series for α and β are identical with the series solution obtained in Chapter II for the function Y, which is the integrand in the exponent of the eikonal solutions of the equation

$$\Psi'' + \lambda^2 k^2 \Psi = 0 \quad . \tag{76}$$

Recalling what we did in Chapter II, if the solution of (1) is represented as

---

* $\phi_{1,\,2n}$ and $\phi_{2,\,2n+1}$ are the two components of the vector $\vec{\phi}_1$ with 2N and 2N+1 backscatterings respectively.

$$\Psi(x) = e^{\lambda\int^{x} y(x')dx'} \tag{77}$$

we obtain the Riccati equation for y

$$y' + \lambda y^2 + \lambda k^2 = 0. \tag{78}$$

If we put $dx_1 = \lambda dx$ we have

$$y' + y^2 + k^2 = 0. \text{ (prime is differentiation w.r.t.x.)} \tag{79}$$

The first approximation for y as before is

$$y = \pm\, ik. \tag{80}$$

To go to the next higher approximation we use the method of quasilinearization. This is explained in Chapter XI in great detail. To give a glimpse into this method we write Equation (79) as

$$y' = -y^2 - k^2 \ ; \qquad y(0) = ik_0 \text{ (a constant)}$$

and replace y^2 by $\max_{y_0} [2y_0y - y_0^2]$ and thus obtain

$$y' = -\max_{y_0}[2y_0y - y_0^2] - k^2$$

$$y' = \min_{y_0}[y_0^2 - 2y_0y - k^2]$$

we take the companion equation

$$y' = y_0^2 - 2y_0y - k^2 \quad \text{with} \quad y(0) = ik_0$$

which is linear in y and solve for it. The solution is better approximation if we start with a reasonable initial approximation for y_0. To obtain the next higher approximation we assume the solution y of the above equation as the initial approximation and continue the process further. This procedure is explained in all its details in Reference [9] and a later chapter. The procedure converges quadratically and fast enough if the initial approximation is near the actual solution. We adopt this procedure here.

The next higher approximation to y if we take $y_0 = ik(x)$ is obtained by solving the linearized equation as shown above,

$$y' = -2k^2 - 2iky \tag{81}$$

with $y(0) = ik_0$ (a constant). Hence,

$$y = -2\int_0^x dx' k^2(x') \exp\left[-2i\int_{x'}^x k(z')\, dz'\right] + y(0) \exp\left[-2i\int_0^x k\, dz\right]. \tag{82}$$

Integrating the first integral on the right-hand side by parts, we have

$$y = ik(x) - ik(0) \exp\left[-2i\int_0^x k(z')\, dz'\right] + y(0) \exp\left[-2i\int_0^x k\, dz\right] - \tag{83}$$

$$- i\int_0^x k'(x')\, dx' \exp\left[-2i\int_{x'}^x k(z)\right] dz .$$

Therefore, the solution is [since $y(0) = y_0(0) = ik(0) = ik_0$ by assumption]

$$y = ik(x) - i\int_0^x k'(x')\, dx' \exp - 2i\int_{x'}^x k(z)\, dz. \tag{84}$$

Integrating the integral on the right-hand side by parts, we again obtain (assuming all the derivatives of $k(0)$ to be equal to zero).

$$y = ik(x) - \frac{k'}{2k} + \frac{1}{2}\int_0^x \left\{\frac{k''}{k} - \frac{k'^2}{k^2}\right\} dx' \exp\left[-2i\int_{x'}^x k(z)\, dz\right] . \tag{85}$$

Remembering that $dx_1 = \lambda dx$, the integrated parts up to the second term turn out as

$$y = ik(x) - \frac{k'}{2k\lambda} \tag{86}$$

and the solution is

$$\Psi \sim \exp\left(\int_0^x \left(i\lambda k - \frac{k'}{2k}\right) dx\right) \tag{87}$$

which is the WKB solution. If we have taken the initial approximation as

$$y = -ik \tag{88}$$

the WKB solution will be

$$\Psi = \exp\left[\int_0^x \left(-i\lambda k - \frac{k'}{2k}\right) dx'\right]. \tag{89}$$

If we continue integrating by parts the right-hand side of Equation (84) we will obtain for the n-th integrated part L_n the following recursive relation, in general.

$$L_n^+ = \frac{1}{2ik}\frac{d}{dx_1}(-L_{n-1}^+) \quad n = 1, 2, \cdots \tag{90}^*$$

$$L_1^+ = + ik \tag{91}$$

and the n-th term for the integrated part of the above solution, starting with y = -ik, is given by

$$L_n^- = -\frac{1}{2ik}\frac{1}{dx_1}(-L_{n-1}^-) \qquad n = 1, 2, \ldots \tag{92}^*$$

$$L_1^- = -ik. \tag{93}$$

The first few terms of the integrand in the eikonal solution are given by the following if $y_0 = + ik$;

$$y^+ = ik - \frac{k'}{2k} + \frac{1}{4ik}\left\{\frac{k''}{k} - \frac{k'^2}{k^2}\right\} + \frac{1}{8k}\left\{\frac{k'''}{k^2} - \frac{4k''k'}{k^3} + \frac{3k'^3}{k^4}\right\} + \cdots . \tag{94}$$

If we start with $y_0 = -ik$, we have the series

$$y^- = -ik - \frac{k'}{2k} - \frac{1}{4ik}\left\{\frac{k''}{k} - \frac{k'^2}{k^2}\right\} + \frac{1}{8k}\left\{\frac{k'''}{k^2} - \frac{4k''k'}{k^3} + \frac{3k'^3}{k^4}\right\} + \\ + \cdots = (y^+)^* . \tag{95}$$

It is to be noted that the $y^{\pm}$ series though very similar to the series for y in Chapter II or V is not identical. This is because this solution sums up a different class of perturbation terms. We shall not go into it now. In each term of Equations (94) and (95) the differentiations are with respect to the variable x_1, and each term brings in a factor of $1/\lambda$,

---

* We start iteration with n = 2 on the R.H.S. of these equations defining $L_1^{\pm}$ ab initio.

and the n-th order term has $1/\lambda^{n-1}$ as a factor. Therefore, let us present the whole solution y^+ and y^- given by

$$y^+ = ik - i\int_0^x dx'k' \exp\left[-2i\int_{x'}^x k(z)\,dz\right] \tag{96}$$

and

$$y^- = -ik + i\int_0^x dx'k' \exp\left[+2i\int_{x'}^x k(z)\,dz\right]. \tag{97}$$

Instead of stopping with the first few terms we assume the solutions as

$$\exp\left[\lambda\int_0^x y\,dx\right] \text{ or } \exp\left[\lambda\int_0^x y^*\,dx\right].$$

The splitting of the wave solution corresponding to the above can be effected by the transformation

$$\begin{bmatrix}\Psi\\ \Psi'\end{bmatrix} = \begin{pmatrix}1 & 1\\ \lambda y & \lambda y^*\end{pmatrix}\begin{bmatrix}u_1\\ v_1\end{bmatrix}. \tag{98}$$

This leads to the transport equations:

$$\frac{d}{dx}\begin{pmatrix}u_1\\ v_1\end{pmatrix} = \begin{bmatrix}Au_1 + Bv_1\\ -Cu_1 - Dv_1\end{bmatrix} \tag{99}$$

where

$$\begin{aligned} A &= \frac{\lambda y' + \lambda^2 y^* y + \lambda^2 k^2}{\lambda(y^* - y)} \\ &= \frac{-2\lambda k(R - iI) + 2\lambda^2 k^2 - 2\lambda^2 kR + \lambda^2(R^2 + I^2)}{-2i\lambda(k - R)} \end{aligned} \tag{100}$$

where

$$R = \int_0^x dx'k'(x')\cos 2\int_{x'}^x k(z)\,dz \tag{101}$$

and

$$I = \int_0^x dx'k'(x') \sin 2 \int_{x'}^{x} k(z)\,dz \qquad (102)$$

$$C = A - \lambda y$$

$$B = \frac{-2\lambda k(R + iI) + \lambda^2\{I^2 - R^2 + 2kR + 2iI(k - R)\}}{-2i\lambda(k - R)} \qquad (103)$$

$$-C = B^* \quad \text{and} \quad -D = A^*. \qquad (104)$$

Once we have the transport equations, the Bremmer series solutions for suitable boundary conditions can be arrived at as explained in earlier sections.

We want to point out at this stage that if we look at the coefficients A, B, C, and D of the tranport equation arrived at in earlier sections, we find that there may be difficluties at the so-called 'turning point', i.e., when $k(x)$ goes to zero in the inhomogeneous medium. Let us now consider the denominator $\lambda(y^* - y)$ of the coefficients of the transport equations in this section.

$$\lambda(y^* - y) = -2i\lambda\left[k(x) - \int_0^x dx'k' \cos 2\left(\int_{x'}^{x} k(z)\,dz\right)\right]. \qquad (105)$$

The right-hand side can be rewritten after partial integration as

$$\lambda(y^* - y) = -2i\lambda\left[k(0) \cos 2\theta(x, 0) + 2\int_0^x k^2 \sin 2\theta(x, x')\,dx\right] \qquad (106)$$

with

$$\theta(x, x') = \int_{x'}^{x} k(z)\,dz. \qquad (107)$$

Let the first turning point occur at $x = x_0$. Let us investigate whether the denominator $\lambda(y^* - y)$ goes to zero at this value of x_0. This can happen only if

$$k(0) = \frac{-2\int_0^x k^2 \sin 2\theta(x_0, x')\,dx'}{\cos 2\theta(x_0; 0)} \qquad (108)$$

Let us take the case when k(0) is positive, i.e., when k starts with a positive value.

Equation (108) can never be satisfied as long as

$$\int_0^{x_0} k(z)\,dz \le \frac{\Pi}{4}\,. \tag{109}$$

Similarly, if k starts with a negative value, i.e., if k(0) is negative and if k(x) goes to zero at $x = x_0$ for the first time, $\sin 2\theta(x_0, x')$ remains negative and $\cos 2\theta(x_0, 0)$ positive and Equation (108) remains unsatisfied.

This conditior (109) is the condition that ensures that we do not meet with the turning point difficulty if there is one turning point in the problem. Thus, the first quasi-linearization approximation to the eikonal enables us to overcome the turning point difficulty under certain conditions. There might be difficulties at other points.

The next approximation in the quasilinearization procedure can yield another type of summation of the eikonal series.

In conclusion, we want to point out that we have presented a unified picture of the various methods of generalizing the Bremmer series relating to the scalar wave equation with and without the inhomogeneous term. We have also shown how by employing the method of quasilinearization an approximation to the eikonal solution which takes into account the effects of the perturbation terms to all orders in a particular way can be obtained. This approximation may prove effective for bypassing the turning point difficulty under certain conditions.

REFERENCES

[1] Bellman, R. and R. Vasudevan, 'On the Generalizations of the Bremmer Series, Solutions of the Wave Equations'. Journal of Mathematical Analysis and Applications 52 (1975), 151.

[2] van Kampen, N.G., 'Higher Corrections to the WKB Approximations', Physica 35 (1967), 70.

[3] Keller, H.B. and J.B. Keller, 'Exponential Like Solutions of Systems of Linear Ordinary Differential Equations', Journal of Soc. Ind. Appl. Math. 10 (1962), 246.

[4] Bahar, E., 'Generalized WKB Method with Applications to Problems of Propagation in Nonhomogeneous Media'. Journal of Math. Physics 8 (1967), 1735.

[5] Chandrasekhar, S., 'Plasma in a Magnetic Field', R.K.M. Landshoff (ed.), Stanford University Press, California, 1958, 14.

[6] Wing, G.M., 'Invariant Imbedding and Generalization of the WKB Method and Bremmer Series', Journal of Math. Analysis and Applications 48 (1974), 400.

[7] Bellman, R. and R. Kalaba, Quasilinearization and Nonlinear Boundary Value Problems, American Elsevier Publishing Co., New York, 1965.

[8] Bellman, R. and R. Vasudevan, 'Wave Equation with Sources, Invariant Imbedding and Bremmer Series Solutions', Journal of Math. Analysis and Applications 48 (1974), 17.

[9] Bellman, R. and R. Kalaba, Quasilinearization and Nonlinear Boundary Value Problems, American Elsevier Publishing Co., New York, 1965.

[10] Bailey, V.A., 'Reflection of Waves by an Inhomogeneous Medium', Phys. Review 96 (1954), 855.

[11] Langer, R.E., Phys. Review 51 (1937), 669.

[12] Miller, S.C., Jr. and R.H. Good, Jr., 'WKB Type Approximations to Schrödinger Equations', Phys. Review 91 (1953), 174.

[13] Walker, L.R. and N. Wax, Journal App. Phys. 17 (1946), 1043.

CHAPTER VII

TIME DEPENDENT PROCESSES

1. INTRODUCTION

In this chapter we will deal with time dependent transport of particles in a medium with given parametric values describing the interaction of the particles with the medium. Imbedding methods to study the emergent beams are dealt with in detail. Laplace transform techniques are also applied to arrive at the solutions both internal and external. When the velocity of the particles become large along with the collision cross section σ in such a way that their ratio is finite, the transport problem in the limit reduces to the diffusion problem [1]. The study of the diffusion equation employing imbedding methods is presented briefly. The inversion of the Laplace transform is a subject which we do not discuss here, but refer the reader to elegant methods developed in Reference [2].

The phenomenon of criticality in a rod in which transport of particles are taking place with production of particles during collisions is analyzed in some detail. The eigen-length problem, and the numerical algorithm to compute the successive critical lengths is studied [3]. The eigen-value problems connected with Sturm-Liouville systems and their solution via imbedding principles are briefly introduced.

The time dependent wave equation is analyzed reducing it to a set of first order partial differential equations. With proper initial conditions Bremmer series like solutions are obtained for the time dependent solutions, which can be easily extended to the case when the wave equation has time dependent source term also [4].

The last section describes the famous Weiner integrals and its relevance to the study of diffusion equations [5], [6]. Derivation of the Schrödinger-like equation from the path integral is briefly outlined [7], [8].

2. TIME DEPENDENT TRANSPORT PROBLEMS

The time dependent transport of particles in a rod is a bit more complicated than the steady state transport considered in detail in Chapter III. If we assume c as the velocity of the particle, and define the density of the right moving particles at the position z, at time t as u(z, t), then cu(z, t) is the flux per unit time that croses the layer z. In Chapter III for the steady state problem we have assumed c to be a unity and therefore ignored it. Let v(z, t) represent the density of the left moving particles. The forward equations for u and v at any point inside the medium can be written by analyzing the physical interactions that take place within region $z - \Delta < z < z + \Delta$, in a time interval $t - \Delta/c \leq t < t + \Delta/c$. As we described in Chapter III, let $\sigma(z, t)\Delta$ be the collision cross section for the medium, f(z, t) the forward scattered fraction and b(z, t) the backward scattered fraction, for both the u and v streams. Hence, after the input-output analysis relating $u(z+\Delta, t+\Delta/c)$ to u(z, t) (the medium does not change due to interaction), we write

$$\frac{\partial u}{\partial z} + \frac{1}{c}\frac{\partial u}{\partial t} = \sigma(z, t)\{(f(z, t)-1)u(z, t)+b(z, t)v(z, t)\}, \tag{1}$$

The corresponding equation for v(z, t) is

$$-\frac{\partial v}{\partial z} + \frac{1}{c}\frac{\partial v}{\partial t} = \sigma(z, t)\{b(z, t)u(z, t)+(f(z, t)-1)v(z, t)\}. \tag{2}$$

More sophisicated versions of the coupled equations for these expected numbers or product density functions u(z, t) and v(z, t) including multicomponent systems have been studied by Ramakrishnan et al. [9]. If we suppose that a single particle was injected at the right end $z = x$, at time $t = 0$, and that both the ends are non-reentrant surfaces we have for the boundary and initial conditions for the partial differential Equations (1) and (2)

$$\begin{aligned} cu(0, t) &= 0 \qquad && t \geq 0 \\ cv(x, t) &= \delta(t) && \\ u(z, 0) &= v(z, 0) = 0 && 0 \leq z < x \end{aligned} \tag{3}$$

$\delta(t)$ is the Dirac delta function.

To simplify the analysis further, if we assume the functions f, b, and σ such that

$$f = b = 1 \text{ and } \sigma = \text{a constant} \tag{4}$$

Equations (1) and (2) reduce to

$$\frac{\partial u}{\partial z} + \frac{1}{c}\frac{\partial u}{\partial t} = \sigma v \tag{5}$$

$$-\frac{\partial v}{\partial z} + \frac{1}{c}\frac{\partial v}{\partial t} = \sigma u \,. \tag{6}$$

This set of Equations coupled with the boundary conditions (3) define the time dependent transport problem and the imbedding analysis can be carried through as in the case of the time independent problems.

To this end we define the reflection as follows: $cR(z, t, \tau)$ = the mean total number of particles emergent from the rod at the point $z = x$, up to time t, due to a single particle incident at $z = x$, at time $t = \tau$, $0 \le \tau \le t$. Let us also assume that $cR(x, t, \tau) = 0$ for $t \le \tau$. Of course, the rate at which particles emerge out of $z = x$, due to unit input, is the differential of the above R with respect to t and we call this the reflected flux r coming out in the interval, t and t + dt

$$cr(x, t, \tau) = c\,\frac{\partial R}{\partial t}\,(x, t\ \tau)\,. \tag{7}$$

By the particle counting process, we can arrive at the differential equation governing $R(x, t, \tau)$ by changing not only the length of the rod x, but also the time points t and τ to allow for the time taken by the particle to traverse the additional distance, moving in either way. Hence, we compute the function $cR(x + \Delta, t + \Delta/c, \tau - \Delta/c)$ by considering a single particle entering at the right end $z = x + \Delta$. In the interval $x \le z < x + \tau$, it may experience a collision with probability $\sigma\Delta$, in which case the contribution to the reflected flux is $\sigma\Delta$. At the same time a fraction $\sigma\Delta$ of the incident beam reached the end of the sub-rod of length x at time τ. If there had been no collision $(1 - \sigma\Delta)$ part of the incident beam would enter the sub-rod. In total, the sub-rod sees an input of $1 + O(\Delta)$ particles at the end x. Since the rate of emergence is $cr(x, t, \tau)$ from the sub-rod allowing for both the possibilities of collision and non-collision in moving to the right through the region z, $x \leq z \leq x + \Delta$, the number coming out at $z = x + \Delta$, up to time t is given by $c\int_{\tau}^{t} r(x, t', \tau)\, dt'$. The particles that emerge in the region z; $x \leq z \leq x + \Delta$ at time t', $0 \leq t' \leq t$, may again be scattered back into the sub-rod and come back as a reflected flux up to time t and this will constitute an amount $c^2\sigma\int_{\tau}^{t} r(x, t', \tau)R(x, t, t')\, dt'$. Combining all these effects and retaining terms of order at most $O(\Delta)$ we obtain

$$cR(x + \Delta, t + \Delta/c, \tau - \Delta/c) = \sigma\Delta + cR(x, t, \tau) + \\ + c^2\sigma\Delta\int_{\tau}^{t} R(x, t, t')r(x, t', \tau)\, dt'. \tag{8}$$

This leads us to

$$\frac{\partial R}{\partial x} + \frac{1}{c}\frac{\partial R}{\partial t} - \frac{1}{c}\frac{\partial R}{\partial \tau} = \frac{\sigma}{c} + \sigma c\int_{\tau}^{t} R(x, t, t')r(x, t', \tau)\, dt'. \tag{9}$$

Since c and σ are constants, we can take R and r as functions which depend on the time difference $t - \tau$. Hence, Equation (9) can be rewritten as

$$\frac{\partial R}{\partial x}(x, T) + \frac{2}{c}\frac{\partial R(x, T)}{\partial T} = \frac{\sigma}{c} + \sigma c\int_{0}^{T} R(x, T - t')r(x, t')\, dt', \tag{10}$$

where $T = t - \tau$ of the previous equation and

$$R(x, T) = 0 \quad T \leq 0, \quad x \geq 0 \tag{11}$$

$$R(0, T) = 0 \quad T \geq 0 \tag{12}$$

The particle counting technique may still be employed even if f, b, and σ are fairly general functions of z and t and if these depend only on z and not on t. The reflection function can always be expressed as a function of the time difference as we expressed in Equation (10). Equations for other emergent functions can be easily deduced. Multi-component particle transport with time dependence can also be similarly handled. Besides the particle counting method, we can also use alternative analytical techniques in the time dependent case. We can write the Riccati transformations in this time dependent case as we did in the time independent case. Let us define for the left moving beam or v beam

$cr_\ell(a, b, t)$ = rate of reflected flux emergence for the left moving v beam at the time interval t and t + dt.

$ct_\ell(a, b, t)$ = rate of transmitted flux emerging for the left going v beam between t and t + dt.

For the above two cases, b is the point of incidence, and a the farther end.

Similarly for the right moving beam, we define $cr_r(a, b, t)$ and $ct_r(a, b, t)$. In these two cases a is the point of incidence and b the farther end. We can now write down the Riccati transformations, relating u and v which we have dealt with in Chapter III (Section 5) in the time independent case:

$$cu(z, t) = c^2 \int_0^t r_\ell(0, z, t - t')v(z, t')\, dt' + c^2 \int_0^t t_r(0, z, t')u(0, t - t')\, dt'. \tag{13}$$

The second term is always zero for the initial conditions we have assumed. The inverse Riccati transformation reads

$$cv(0, t) = c^2 \int_0^t t_\ell(0, z, t - t')v(z, t')\, dt' + c^2 \int_0^t r_r(0, z, t')u(0, t - t')\, dt'. \tag{14}$$

Of course, both u and v should be written as functions of z, t and x, the length of the medium, i.e., we should write u(z, t, x) and v(z, t, x), r_ℓ, r_r and t_ℓ, t_r, as also functions of the length in which the imbedding takes place, i.e., $r_\ell(0, z, t - t')$ refers to the reflection from that part of the rod extending from 0 to z, i.e., of length z, similar explanations hold for other functions t_ℓ, r_r and t_r.

If we now apply the operator $(\partial/\partial z + 1/c\ \partial/\partial t)$ to the left-hand side of Equation (13) we obtain from Equation (5) and (6)

$$\sigma v(z, t) = c \int_0^t \left[\frac{\partial r_\ell}{\partial z}(0, z, t - t') + \frac{1}{c}\frac{\partial r}{\partial t}(0, z, t - t')\right] v(z, t')\, dt' + c \int_0^t r_\ell(0, z, t - t') \frac{\partial v}{\partial z}(z, t')\, dt' + r_\ell(0, z, 0) v(z, t). \tag{15}$$

In writing (15) we recall that u(z, 0) = v(z, 0) = 0 due to our initial conditions. If now we substitute for $\partial v/\partial z$ from Equation (6) to evaluate the third term on the right-hand side of (15) and do a partial integration over t' we obtain

$$[\sigma - 2r_\ell(0, z, 0)]v(z, t) = c \int_0^t \left\{\frac{\partial r_\ell(0, z, t - t')}{\partial z} + \frac{1}{c}\left(\frac{\partial}{\partial t} - \frac{\partial}{\partial t'}\right) r_\ell(0, z, t - t')\right\} v(z, t')\, dt' - c\sigma \int_0^t r_\ell(0, z, t - s) u(z, s)\, ds. \tag{16}$$

We now substitute in the last term on the right-hand side of Equation (16) for u(z, s) as

$$u(z, s) = c \int_0^s r_\ell(0, z, s - t') v(z, t')\, dt' \tag{17}$$

and change the order of integration to obtain

$$v(z, t)[\sigma - 2r_\ell(0, z, 0)] = c \int_0^t \left\{\left[\frac{\partial}{\partial z} + \frac{1}{c}\left(\frac{\partial}{\partial t} - \frac{\partial}{\partial t'}\right)\right] r_\ell(0, z, t - t')\right\} v(z, t')\, dt' - \sigma c^2 \int_0^t v(z, t')\, dt' \int_0^{t-t'} r_\ell(0, z, t - t' - y) r_\ell(0, z, y)\, dy \,. \tag{18}$$

If in the above we assume for a moment that $r_\ell(0, z, 0) = \sigma/2$, we obtain the equation satisfied by $r_\ell(0, z, T)$ as

$$\frac{\partial r_\ell(0, z, T)}{\partial z} + \frac{2}{c}\frac{\partial r_\ell}{\partial T}(0, z, T) = \sigma c \int_0^T r_\ell(0, z, T - y) r_\ell(0, z, y)\, dy. \tag{19}$$

The assumption that $r_\ell(0, z, 0) = \sigma/2$ is consistent with our Equation (10) governing $R_\ell(X; T)$ and the initial conditions contained in (11) and (12). If in Equation (10) we put $T = 0$ we obtain

$$\left.\frac{\partial R(X, T)}{\partial T}\right|_{T=0} = \sigma/2 = r_\ell(0, x, T)\Big|_{T=0} \,. \tag{20}$$

We can also arrive at Equation (10) by the Riccati transformation, starting with the equations for quantities

$$c\bar{u}(z, t) = c \int_0^t u(z, t')\, dt' \tag{21}$$

$$c\bar{v}(z, t) = c \int_0^t v(z, t')\, dt' \tag{22}$$

which satisfy the coupled set of differential equations

$$\frac{\partial \bar{u}}{\partial z}(z, t) + \frac{1}{c}\frac{\partial \bar{u}}{\partial t}(z, t) = \sigma\bar{v}(z, t) \tag{23}$$

$$-\frac{\partial \bar{v}(z, t)}{\partial z} + \frac{1}{c}\frac{\partial \bar{v}}{\partial t}(z, t) = \sigma\bar{u}(z, t) \tag{24}$$

$$c\bar{u}(0, t) = 0 \qquad t \geq 0 \tag{25}$$

$$c\bar{v}(x, t) = 1 \qquad t > 0 \tag{26}$$

$$c\bar{u}(z, 0) = c\bar{v}(z, 0) = 0 \qquad 0 \leq z < x$$

and

$$cR(x, t) = c\bar{u}(x, t). \tag{27}$$

The transformation relating $\bar{u}$ to v is given by

$$c\bar{u}(z, t) = c^2 \int_0^t R_\ell(0, z, t - t')v(z, t')\, dt'. \tag{28}$$

If we go through the analysis that was done for deriving the equation for $r_\ell(0, z, t - t')$ we obtain utilizing the initial condition $R_\ell(0, z, 0) = 0$ the same equation as (10) which was obtained by particle counting techniques. Considerable analysis has been carried out to explain the initial value for $r_\ell(0, X, T)$ obtained in Equation (20) which seems to be a strange result [1]. Hoever, if we write down the equation for $r_\ell(0, X, T)$ by imbedding methods for an input of $cv(X, T) = \delta(t)$, at the end $Z = X$, we can after a careful analysis obtain the same result for the rate of emergence.

If we want to reduce the transport equations for the time dependent case so as to look similar to the equations obtained in the time independent case we use the technique of Laplace transformation and transform the time variable from the equations. To this end we define

$$L_T\{c\bar{u}(z, T)\} = \int_0^\infty e^{-sT}\, c\bar{u}(z, T)\, dT = \bar{U}(z, s) \tag{29}$$

$$L_T\{c\bar{v}(z, T)\} = \int_0^\infty e^{-sT}\, c\bar{v}(z, T)\, dT = \bar{V}(z, s). \tag{30}$$

We then find that Equations (23) and (24) become

$$\frac{d\bar{U}(z, s)}{dz} = -\frac{s}{c}\bar{U}(z, s) + \sigma\bar{V}(z, s) \tag{31}$$

$$-\frac{d\bar{V}(z, s)}{dz} = \sigma\bar{U}(z, s) - \frac{s}{c}\bar{V}(z, s) \tag{32}$$

with the boundary conditions corresponding to (25) and (26) given by

$$\bar{U}(0, s) = 0 \tag{33}$$

$$\bar{V}(x, s) = \frac{1}{s} \tag{34}$$

By applying the familiar method of finding the $\tilde{r}(x, s)$, the reflection from the medium, we arrive at the imbedding Equation (19) for $r(0, z, t)$

$$\frac{d}{dx}\tilde{r}(x, s) = \sigma - \frac{2s}{c}\tilde{r}(x, s) + \sigma\tilde{r}^2(x, s), \tag{35}$$

where $\tilde{r}(x, s) = L_T\{cr(x, t)\}$

The Laplace transform of the reflected flux up to time t, in view of the boundary condition (34), is given by

$$\bar{U}(x, s) = \frac{1}{s}\tilde{r}(x, s) = cR(x, s), \text{ where } R(x, s) = L_T R(x, t) \tag{36}$$

We can now write the differential equations for $\bar{U}(x, s)$ itself. We will call $\bar{U}(x, s)$ as $cR(x, s)$ and we obtain from (35)

$$\frac{\partial R(x, s)}{\partial x} = \frac{\sigma}{sc} - \frac{2s}{c}R(x, s) + c\sigma sR^2(x, s) . \tag{37}$$

By inverting this equation, we obtain the original Equation (10). In view of the boundary condition $c\bar{u}(z, 0) = c\bar{v}(z, 0) = 0$ $(0 \le z \le x)$, we have

$$\frac{\partial R(x, T)}{\partial x} + \frac{2}{c}\frac{\partial R}{\partial T}(x, T) = \frac{\sigma}{c} + \sigma c\int_0^T \frac{\partial R}{\partial t'}(x, t')R(x, t-t')\,dt'. \tag{38}$$

If instead of the condition $cv(x, t) = \delta(t)$ in Equation (3), we have a time dependent incidence given by

$cv(x, t) = f(t)$ in the transformed version of the problem we will have

$$\bar{V}(x, s) = L_T\left\{ \int_0^T f(t')\, dt' \right\}. \tag{39}$$

Hence, due to the linearity of the equations by superposition principle we will have for $\bar{U}(x, s)$ the result

$$\bar{U}(x, s) = sF(s)R(x, s)\,; \qquad F(t) = \int_0^T f(t')\, dt'. \tag{40}$$

Inverting this we have for the reflected flux up to time T and this can easily be found to be

$$c\bar{u}(x, T) = c\int_0^T f(t)R(x, T - t)\, dt. \tag{41}$$

Equation (41) is the statement of the Duhamel's principle for the problem on hand and $cR(x, t)$ can be called the unit response function which is evident from its definition.

3. TRANSPORT EQUATION IN THE LIMIT OF LARGE VELOCITIES AND LARGE σ

It is of interest to discuss what happens to the transport equation as the velocity of the particles is allowed to become arbitrarily large with a corresponding increase in the probability of collision. This is physically important since under the assumption of high velocity and small mean free path [10] we arrive at an equation describing a diffusion process or the heat equation. From another direction the discrete random walk process yields diffusion equations in the limit [11]. This observation has been made the basis for a considerable amount of analytic and computational effort centered about the theme of 'Monte Carlo' techniques.

Let us consider Equations (1 and 2) describing the transport of particles in a one-dimensional rod, travelling with constant velocity c, and collision cross section σ assumed constant. Let us assume $f = b = k$ a constant in Equations (1) and (2). Then we have after taking the Laplace transform with respect to the time variable

$$\frac{du(z, s)}{dz} + \frac{s}{c} u(z, s) = \sigma(k-1)u(z, s) + \sigma k v(z, s), \tag{42}$$

$$-\frac{dv(z, s)}{dz} + \frac{s}{c} v(z, s) = \sigma k u(z, s) + \sigma(k-1)v(z, s), \tag{43}$$

with the boundary conditions

$$u(0, s) = 0 \qquad v(x, s) = q(s). \tag{44}$$

We have used for the boundary value of v(x, t) a function f(t) whose Laplace transform is q(s). The above equation can be put into the form

$$\frac{d}{dz}\begin{bmatrix} u(z, s) \\ v(z, s) \end{bmatrix} = \begin{bmatrix} -\frac{s}{c} + \sigma(k-1) & \sigma k \\ -\sigma k & \frac{s}{c} - \sigma(k-1) \end{bmatrix} \begin{bmatrix} u \\ v \end{bmatrix}. \tag{45}$$

Since the coefficients are constants, the matrix can easily be diagonalized and the solution can be obtained after imposing the boundary conditions. We have, after some calculations,

$$u(z, s) = \frac{k\sigma q \sin h\,\lambda z}{\lambda \cos h\,\lambda x + (\frac{s}{c} + (1-k)\sigma) \sin h\,\lambda x} \tag{46}$$

$$v(z, s) = \frac{q(s)\{\sigma\lambda \cos h\lambda z + [\frac{s}{c} + (1-k)\sigma] \sin h\,\lambda z\}}{\lambda \cos h\,\lambda x + [\frac{s}{c} + (1-k)\sigma] \sin h\lambda x} \tag{47}$$

where

$$\lambda^2 = (\frac{s}{c})^2 + \frac{2(1-k)\sigma}{c} s + \sigma^2(1-2k) \tag{48}$$

and the λ in (46) and (47) is $|\lambda|$. We can get rid of the σ^2 term in (48) by taking k = 1/2. This assumption does not impair the generality of the treatment. Other situations with different values of k can be studied by the reader.

Now if we subtract Equation (43) from (42) we obtain, after taking the Laplace inverse

$$\frac{\partial}{\partial z}(u+v) + \frac{1}{c}\frac{\partial}{\partial t}(u-v) = -\sigma(u-v) \; . \tag{49}$$

When c goes to ∞, we have the Fick's law of diffusion

$$\frac{\partial}{\partial z}(u+v) = -\sigma(u-v) \; . \tag{50}$$

Equation (50) represents the statement of Fick's law that the net flux is proportional to the gradient of the concentration in the opposite direction. Since we know the solutions u and v from Equations (46) and (47), let us compute the quantity $\sigma(u-v)$. Let c and σ tend to ∞ uniformly such that c/σ tends to a constant D. Then we find from the equation above the Laplace transform OE $\sigma(u-v)$ as:

$$\sigma(u-v) = F_0(z, s) = \frac{-2q(s)\sqrt{\frac{s}{D}}\cosh\left(z\sqrt{\frac{s}{D}}\right)}{\sinh\left(x\sqrt{\frac{s}{D}}\right)} \tag{51}$$

where we have taken the limits

$$\lambda\Big|_{\substack{c\to\infty \\ \sigma\to\infty}} = \sqrt{\frac{s}{D}} \quad \text{with} \quad \frac{c}{\sigma}\Big|_{\lim} = D \; . \tag{52}$$

Let us add the two Equations (42) and (43) for the value of $k = 1/2$ and then apply the limits

$$\begin{aligned} s(u+v)\Big|_{\substack{\lim \\ c\to\infty \\ \sigma\to\infty}} &= -\frac{c}{\sigma}\frac{d}{dz}\{\sigma(u-v)\} \\ &= -D\frac{d}{dz}F_0(z, s) \\ &= \frac{s2q(s)\sinh\left(\sqrt{\frac{s}{D}}\,z\right)}{\sinh\left(x\sqrt{\frac{s}{D}}\right)} \; . \end{aligned} \tag{53}$$

Hence, the Laplace transform of $u(z, t) + v(z, t)$ turns out to be

$$(u+v) = \theta(z, s) = \frac{2q(s) \sinh (\sqrt{\frac{s}{D}}\, z)}{\sinh (\sqrt{\frac{s}{D}}\, x)} \tag{54}$$

and

$$\theta(0, s) = 0$$

$$\theta(x, s) = 2q(s) \; . \tag{55}$$

We easily see from Equations (54) and (53) that

$$D \frac{\partial^2 (u+v)}{\partial z^2} = \frac{\partial}{\partial t} (u+v); \qquad \theta(z, t) = (u+v) \tag{56}$$

Here we remember the initial conditions on u(z, t) and v(z, t) at time t = 0 (Equation (3)). Thus original transport Equations (1) and (2) in the limit of large velocity and large collision cross section σ, has become the diffusion Equation (56) with boundary conditions

$$\theta(0, t) = 0; \qquad \theta(x, t) = 2q(t) \quad \text{and} \quad \theta(z, 0) = 0 \; . \tag{57}$$

Thus to analyze the solutions of the diffusion equation we can go to the equivalent transport process of finite c velocity and finite σ, and go over to the limit.

Many physical phenomena, like the flow of heat in a material body, or slowing down of neutron in a medium, or radiation transport, can be represented by the diffusion equation. We will consider the diffusion equation of the type

$$\frac{d\theta}{dt} = D \frac{d^2\theta}{dz^2} \; . \tag{58}$$

We can imagine θ(z, t) as the temperature at any point z of the rod (a one-dimensional version of the rod) at time t, and at the left end the temperature is kept constant. We can specify the flux of heat at the right end as a function of time. Hence, the boundary and initial conditions of this problem are taken as

$$\theta(z, 0) = 0 \quad 0 \le z \le x$$

$$\theta(0, t) = 0 \quad 0 \le t \tag{59}$$

$$-\frac{d\theta}{dz}(z, t)\Big|_{z=x} = f(t), \quad 0 \le t.$$

To compute $\theta(x, t)$ at $z = x$ at any time t, we first convert Equation (58) by the Laplace transform procedure into a second order equation in one variable only, s the Laplace transform variable being a parameter

$$D\frac{d^2\theta(z, s)}{dz^2} - s\theta(z, s) = 0$$

$$\theta(z, s) = \int_0^\infty \theta(z, t)e^{-st}\, ds \,. \tag{60}$$

Since D is a constant, we can use proper units so that it can be reduced to unity. The boundary conditions are

$$\theta(0, s) = 0 \,; \quad -\theta'(x, s) = f(s) = \int_0^\infty f(t)e^{-st}\, dt \,. \tag{61}$$

We then set $\theta(z, s) = u(z, s)$ and $\theta'(z, s) = v(z, s)$ and obtain the coupled set of equations as in particle transport

$$\frac{du(z, s)}{dz} = v(z, s) \tag{62}$$

$$-\frac{dv(z, s)}{dz} = -su(z, s) \tag{63}$$

$$u(0, s) = 0; \quad v(x, s) = -f(s) \,. \tag{64}$$

The reflection function $r(x, s)$, i.e., the unit response function for the system is governed by the equation

$$\frac{\partial r(x, s)}{\partial x} = 1 - sr^2(x, s) \tag{65}$$

with the initial condition

$$r(0, s) = 0 \,. \tag{66}$$

The resulting return value of $u(x, s)$ at $z = x$ can be obtained from the solution of (65), since for the actual input of $v(x, s) = - f(s)$; $u(x, s)$ by Duhamel principle is

$$u(x, s) = - f(s)r(x, s) . \tag{67}$$

The Riccati Equation (65) can be easily solved to yield

$$r(x, s) = \frac{1}{\sqrt{s}} \tanh (x \sqrt{s}) . \tag{68}$$

Thus the temperature at $z = x$ at any time $t > 0$ is found by means of a convolution

$$\theta(x, t) = - \int_0^t dt' f(t-t') L_s^{-1} \left[\frac{\tanh (x \sqrt{s})}{\sqrt{s}} \right] . \tag{69}$$

The Laplace inversion of $[\tanh (x \sqrt{s})/\sqrt{s}]$ can be found from the tables in [12]. We know

$$L_s^{-1} \left(\frac{-1}{\sqrt{s}} \tanh \sqrt{s} \right) = \theta_1 \left(\frac{1}{2}, t \right) ,$$

where θ_1 is the theta function of index one. We also know that if

$$g(s) = L_s [G(t)] \tag{71}$$

$$g(ks) = \frac{1}{k} L_s [G(\frac{t}{k})] . \tag{72}$$

Hence, we have

$$\theta(x, t) = \frac{1}{x} \int_0^t \theta_1 \left(\frac{1}{2}, \frac{t}{x^2}\right) f(t-t') dt' . \tag{73}$$

To find the internal solutions, we can write down the equation for the transmission function and get algebraic equations for $u(z, t)$ and $v(z, t)$ and solve them as indicated in Chapter III. The equations for r and t are initial value equations and they can be solved for all values of z in the interval [0, x] even if the diffusivity D is a function of z. As explained in Chapter III, once r_ℓ, r_r, t_ℓ and t_r are evaluated for various values of their arguments in [0, x] all types of problems with different boundary values and for different lengths of the medium can easily be studied.

4. THE EIGENVALUE PROBLEMS*

The method of invariant imbedding can be very useful in a practical way in solving eigenvalue problems, without worrying about the calculation of the entire set of solutions to the problem. For example, if we start with a steady state problem of particle transport, assuming as we have done earlier that $f = b = 1$ in Equations (1) and (2) of Chapter III, we obtain the transport equations

$$\frac{du}{dz} = \sigma v \tag{74}$$

$$-\frac{dv}{dz} = \sigma u \qquad 0 \leq z \leq x \ . \tag{75}$$

If the boundary conditions are such that

$$u(0) = 0 \quad \text{and} \quad v(x) = 1 \tag{76}$$

the problem represents the situation in which a neutron beam of unit flux enters at $z = x$ and σ is the collision cross section. At each collision the neutron interacting with the medium is replaced by one right-moving and another left-moving particle at the point as a result of the collision. We have assumed the velocity of the particle to be unity and we are interested in the reflected flux of neutrons coming out at the end $z = x$.

The solutions of the above system are easily found to be

$$u(z) = \frac{\sin \sigma z}{\cos \sigma x} \tag{77}$$

$$v(z) = \frac{\cos \sigma z}{\cos \sigma x} \ . \tag{78}$$

But we also know that the denominator will go to zero at values of $x = (n + \frac{1}{2})\pi/\sigma$. Hence, the solutions are valid in the range of $[0, x]$ when $x < (\frac{1}{2})\pi/\sigma$, that is, the rod in which the interactions make the neutron multiply themselves will become critical when x reaches a value equal to $\pi/2\sigma$.

---

* Sections (4) and (5) are based on the treatment by Scott [3].

If we now write dwon the equation for the reflection function r(x) it satisfies the Riccati equation

$$\frac{dr(x)}{dx} = \sigma(1 + r^2) \tag{79}$$

with the initial condition

$$r(0) = 0 \tag{80}$$

and r(x) blows up at the value $x_{cr} = (1/2)\pi/\sigma$ which may be called the first critical point. Hence, the method of finding the first critical length $x_{1,cr}$ is to integrate Equation (79) until it becomes infinite. Let us considerthe usual set of transport equations with coefficients depending on the variable z $(0 \leq z \leq x)$. In general, they can be written down as

$$\frac{du}{dz} = a(z)u(z) + b(z)v(z) \tag{81}$$

$$-\frac{dv}{dz} = c(z)u(z) + d(z)v(z) \; . \tag{82}$$

If on the average more than one particle is produced at each collision, as we have seen in the simple example given earlier, the rod becomes critical for a certain value $x_{1,cr}$ of its length. This depends on the value of the parameters of the problem. At the critical stage no external input is necessary for supporting the population (u, v). Seen from this point of view, we are to solve the set (81) and (82) under the boundary conditions

$$u(0) = 0 \qquad v(x_{cr}) = 0 \; . \tag{83}$$

Provided the coefficients a, b, c, and d are well behaved, this phenomenon of criticality will reoccur at other values of x at $x_{1,cr} < x_{2,cr} < x_{3,cr} \cdots$ etc.

As we already know, u(z) can be related to the v flux by the response function r(z) as

$$u(z) = r(z)v(z) \; . \tag{84}$$

We have also seen already that r(z) will satisfy a differential equation

$$\frac{dr}{dz} = b(z) + (a + d)r + c(z)r^2 \tag{85}$$

and for the relation (84) to hold at z = 0, we should write

$$r(0) = 0 \; . \tag{86}$$

Because if u(z) = 0, and r(z) is finite, v(z) has to be zero at z. If u(z) and v(z) are simultaneously zero at any point, by uniqueness, u(z) and v(z) will be zero throughout the interval. That means we have only the trivial solution for the problem. v(z) is to be non-zero for some interval of length to the right of zero. If v(z) is to be zero, r(z) should be undefined at that point. If not, u(z) will be zero and uniqueness rules this out. In fact, the solution fails at this point. It is this first point $x_{1,cr}$ at which r(z) blows up and v(z) becomes zero for the first time. Thus the method of finding the first critical point is to find the point z = x at which r(z) blows up. Having found this, how do we proceed to find the next critical point. To accomplish this we consider the function

$$S(z) = \frac{1}{r(z)} \tag{87}$$

which does not blow up at this point, but is really zero

$$v(z) = S(z)u(z) \; , \tag{88}$$

and, making use of (88), we can find that S also satisfies a Riccati equation

$$-S'(z) = c(z) + [a + d]S + bS^2 \; . \tag{89}$$

To integrate this equation we start with a value of S(z) at a point z_1, z_1 being less than $x_{1,cr}$, such that the initial value for the equation is given by

$$S(z_1) = \frac{1}{r(z_1)}, \qquad 0 < z_1 > x_{1,cr} \; . \tag{90}$$

As the integration approaches the next critical value (x_1) critical for this equation, u(z) will approach zero and S(z) will tend to become large. At this stage it is necessary to switch back to the r equation and carry on the integration through this critical value $z = \bar{x}_{1,cr}$. This scheme works out very well. The zeroes of the S function will furnish the successive critical points whose distances correspond to the successive eigenlengths of the problem.

5. EIGENVALUE PROBLEMS OF STURM-LIOUVILLE SYSTEMS

Since we can always convert a second order equation of the type

$$y'' + ay' + by = 0 \tag{91}$$

With homogeneous boundary conditions into a set of coupled equations for u = y and v = y' we shall be interested in eigenvalue problems for Sturm-Liouville systems. They can be generally written down as

$$\frac{d}{dz}\left\{k(z, \lambda)\,\frac{du}{dz}\right\} + g(z, \lambda)u(z, \lambda) = 0 \tag{92}$$

with boundary conditions,

$$\left.\begin{aligned} a_1(\lambda)u(0) + b_1(\lambda)u'(0) &= 0 \\ c_1(\lambda)u(x) + d_1(\lambda)u'(x) &= 0 \end{aligned}\right\} \quad 0 \le z \le x\ .$$

We can consider x fixed and λ as a parameter. It is well known that for suitable functions k and g, $a_1(\lambda)$, $b_1(\lambda)$, $c_1(\lambda)$ and $d_1(\lambda)$, etc., one can obtain a discrete set of eigenvalues λ for the system provided $k(z, \lambda) \neq 0$, in the entire interval [0, x].

The system of Equations (92) can be represented by a coupled set of equations for u and v as given below

$$\frac{du}{dz} = \frac{1}{k(z, \lambda)}\, v \tag{93}$$

$$-\frac{dv}{dz} = g(z, \lambda)u(z)\ . \tag{94}$$

Since we are only interested in the eigenvalues, we shall consider the reflection or response function r(z)

$$r(z) = \frac{u(z)}{v(z)} = \frac{u(z)}{k(z, \lambda)u'(z)} \tag{95}$$

or its reciprocal S(z)

$$S(z) = \frac{k(z, \lambda)u'}{u} . \tag{96}$$

r(z) and S(z) satisfy the following nonlinear equations

$$\frac{dr}{dz} = \frac{1}{k(z, \lambda)} + g(z, \lambda)r^2(z) \tag{97}$$

$$-\frac{dS}{dz} = g(z, \lambda) + \frac{1}{k(z, \lambda)} S^2(z) . \tag{98}$$

The boundary conditions on r(z) derived from the boundary conditions on u and v are given by

$$r(0) = \frac{u(0)}{k(0, \lambda)u'(0)} = \frac{-b_1(\lambda)}{k(0, \lambda)a_1(\lambda)} \tag{99}$$

$$r(x) = \frac{-d_1(\lambda)}{k(x, \lambda)c_1(\lambda)} . \tag{100}$$

Suppose we have to determine the eigenvalue for a given extension x of the medium. We assume that a non-trivial solution exists for that interval and start integrating Equation (97) with the initial value for r(0) given by Equation (99). Determination of the eigenvalue reduces to solving the non-linear algebraic equation for λ in the equation after computing r(x).

$$r(x) = \frac{-d_1(\lambda)}{k(x, \lambda)c_1(\lambda)} . \tag{101}$$

The numerical method is to take an initial guess for λ, integrate the r equation up to x and check whether Equation (101) is satisfied. If not satisfied, the value of λ is adjusted, and the process repeated until a value for λ is found which fulfils the initial value condition for r(0), Equation (99) and also the boundary condition (100) at z = x.

If, while integrating $r(z)$ out to $z = x$, we meet with a blow up of its values, we can switch to Equation (98) for $S(z)$ and continue the integration through that point as described earlier. The advantages of adopting this technique for the determination of eigenvalues are discussed in References [3] and [13]. This method of finding the eigenvalues is closely related to a method using the Prüfer transformation [14]. In this method two functions $\varphi(z)$ and $\theta(z)$ are defined by the relations

$$u(z) = \varphi(z) \sin \theta(z) \tag{102}$$

$$k(z, \lambda)u'(z) = \varphi(z) \cos \theta(z) \ . \tag{103}$$

Substituting those into the second order Equation (92), we arrive at

$$\theta'(z) = \frac{1}{k(z, \lambda)} \cos^2 \theta(z) + g(z, \lambda) \sin^2 \theta(z) \ . \tag{104}$$

Since $\tan \theta(z) = u(z)/k(z, \lambda)u'(z)$, the boundary conditions on θ are given by

$$\theta(0) = \omega_1 \quad \text{and} \quad \theta(x) = \omega_2 + n\pi \quad n = 0, 1, 2, \cdots, \tag{105}$$

where

$$\tan \omega_1 = \frac{-b_1(\lambda)}{k(0, \lambda)a_1(\lambda)} \tag{106}$$

and

$$\tan \omega_2 = \frac{-d_1(\lambda)}{k(x, \lambda)c_1(\lambda)} \ . \tag{107}$$

It is evident from Equations (102) and (103) that the response function $r(s) = \tan \theta$ and if Equation (104) is divided by $\cos^2\theta$ we obtain the equation satisfied by r. Similarly, $S = \cot \theta$, and the differential equation for S is arrived at on dividing the whole Equation (104) by $\sin^2\theta$. Thus the two approaches for determining λ are equivalent. However, the imbedding approach can easily be generalized to handle systems of equations or in other words to handle higher order systems quite easily, while the generalization of Prüfer's transformation technique is not very clear, though matrix analogues do exist. Moreover, integration of

$r(x)$ and $S(x)$ are numerically easier simply because there are fewer function evaluations involved. Other advantages are discussed in Reference [13].

6. TIME DEPENDENT WAVE EQUATION

In many fields of physics and engineering like electromagnetic theory, acoustics, or transmission line theory, we deal with partial differential equations of the type

$$\frac{\partial^2 \Psi}{\partial x^2} = k^2(x) \frac{\partial^2 \Psi}{\partial t^2} . \tag{108}$$

Assuming a steady state, we take monochromatic solutions of the wave function

$$\Psi(x, t) = \bar{\Psi}(x)e^{i\omega t} , \tag{109}$$

where ω is the frequency of the wave, and $\bar{\Psi}$ satisfies the equation

$$\frac{\partial^2 \bar{\Psi}}{\partial x^2} + \omega^2 k^2(z)\bar{\Psi}(z) = 0 . \tag{110}$$

The solutions $\bar{\Psi}$ of the equations of the type given by Equation (110) have been studied in previous chapters. If the boundary conditions are time dependent or if k is a function of both x and t, or t only, we have to solve the time dependent wave equation, for example, the initial input may be a pulse at one end of the medium. Can we now develop a Bremmer series type of solution for $\Psi(x, t)$? To study this, let us consider an initial value problem defined by

$$\Psi_{xx} - k^2\Psi_{tt} = 0 \tag{111}$$

$$\Psi(x, 0) = \Psi^0(x) \tag{112}$$

$$\Psi_t(x, 0) = \Psi_t^0(x) . \tag{113}$$

We assume that the functions given here are continuously differentiable, and that k satisfies the required conditions for effecting the Liouville transformation on Ψ and x. Let us now write down Equation (111) as

$$\frac{d}{dx}\begin{pmatrix}\Psi_t\\ \Psi_x\end{pmatrix} = \begin{pmatrix}0 & 1\\ k^2 & 0\end{pmatrix}\frac{d}{dt}\begin{pmatrix}\Psi_t\\ \Psi_x\end{pmatrix} \tag{114}$$

$$\vec{\Psi}_x = \underset{\sim}{A}\,\vec{\Psi}_t \quad \text{with } \vec{\Psi} = \begin{pmatrix}\Psi_t\\ \Psi_x\end{pmatrix}. \tag{115}$$

The suffixes refer to the variable with respect to which the vector $\vec{\Psi}$ is being differentiated.

The matrix A can be diagonalized by the matrix

$$P = \begin{pmatrix}1 & 1\\ k(x) & -k(x)\end{pmatrix} \tag{116}$$

and if we make the transformation of the vector $\vec{\Psi}$ by

$$\vec{\Psi} = P\,\vec{\tilde{\varphi}} \tag{117}$$

Equation (114) becomes

$$\tilde{\varphi}_{1,x} - k\tilde{\varphi}_{1,t} + \frac{k'}{2k}\tilde{\varphi}_1 = +\frac{k'}{2k}\tilde{\varphi}_2 \tag{118}$$

$$\tilde{\varphi}_{2,x} + k\tilde{\varphi}_{2,t} + \frac{k'}{2k}\tilde{\varphi}_2 = +\frac{k'}{2k}\tilde{\varphi}_1 \tag{119}$$

where $\vec{\tilde{\varphi}} = \begin{bmatrix}\tilde{\varphi}_1\\ \tilde{\varphi}_2\end{bmatrix}$.

The above system of equations can be written compactly as

$$\vec{\tilde{\varphi}}_x = k\sigma_z\vec{\tilde{\varphi}}_t - \frac{k'}{2k}\vec{\tilde{\varphi}} + \frac{k'}{2k}\sigma_x\vec{\tilde{\varphi}}\,. \tag{120}$$

If we now make the following transformations

$$\vec{\varphi} = k^{1/2}\,\vec{\tilde{\varphi}} \tag{121}$$

and $dz = k\,dx$, we obtain

$$\frac{d}{dz}\vec{\varphi} = \sigma_z\vec{\varphi}_t + \frac{k'}{2k}\sigma_x\vec{\varphi} \tag{122}$$

or equivalently

$$\varphi_{1,z} - \varphi_{1,t} = +\frac{k'}{2k}\varphi_2 \tag{123}$$

$$\varphi_{2,z} + \varphi_{2,t} = +\frac{k'}{2k}\varphi_1 \ . \tag{124}$$

If we now ignore the coupling between the two components we see that the two WKB solutions are of the form

$$\begin{aligned}\varphi_1 &\sim \varphi_1(z + t)\\ \varphi_2 &\sim \varphi_2(z - t) \ .\end{aligned} \tag{125}$$

If we consider Equation (122) and differentiate it once again with respect to z, we have

$$\frac{d^2\vec{\varphi}}{dz^2} = \sigma_z\vec{\varphi}_{t,z} + \frac{k'}{2k}\sigma_x\vec{\varphi}_z + \left(\frac{k'}{2k}\right)'\sigma_x\vec{\varphi} \ , \tag{126}$$

substituting for $\vec{\varphi}_{tz}$ as

$$\vec{\varphi}_{t,z} = \left[\sigma_z\vec{\varphi}_t + \frac{k'}{2k}\sigma_x\vec{\varphi}\right]_t \tag{127}$$

and summing up the components of the two equations we arrive at

$$\varphi^s_{zz} - \varphi^s_{tt} = \left[\frac{k'^2}{4k^2} + \left(\frac{k'}{2k}\right)'\right]\varphi^s = \left(\frac{k''}{2k} - \frac{1}{4}\frac{k'^2}{k^2}\right)\varphi^s \tag{128}$$

which is a scalar equation for the sum of the two components

$$\varphi^s = \varphi_1 + \varphi_2 \ . \tag{129}$$

This equation could also be directly obtained from Equation (111) if we make the Liouville transformation

$$\varphi^{s} = k^{+1/2}\psi \quad \text{or} \quad \Psi = k^{-1/2}\varphi^{s} \tag{130}$$

$$dz = k\,dx\ . \tag{131}$$

In Equation (128) all the differentiations are with respect to the new independent variable z. The initial values for Equation (128) corresponding to the initial value (112) and (113) are

$$\varphi^{S}(z,\ 0) = \varphi^{0}(z) \tag{132}$$

$$\varphi_t^{S}(z,\ 0) = \varphi_t^{0}(z)\ . \tag{133}$$

We can derive the initial conditions for φ_1 and φ_2 from the initial conditions (132) and (133) and the differential equations (123) and (124) themselves. After some manipulations, we arrive at the initial conditions for $\varphi_1(z,\ t)$ and $\varphi_2(z,\ t)$ as:

$$\varphi_1(z,\ 0) = \frac{1}{2}\,\varphi^{0}(z) + \frac{1}{2}k^{-1/2}(z)\int_0^z \varphi_t^{0}(z')k^{+1/2}(z')\,dz' + + ak^{-1/2}(z) \tag{134}$$

and

$$\varphi_2(z,\ 0) = \frac{1}{2}\,\varphi^{0}(z) - \frac{1}{2}k^{-1/2}\int_0^z \varphi_t^{0}(z')k^{+1/2}(z')\,dz' - - ak^{-1/2}(z)\ , \tag{135}$$

where a is the integration constant.

Let us now consider the partial differential equations (123) and (124). The characteristics of Equation (123) are

$$z + t = \text{constant}, \tag{136}$$

and that of Equation (124)

$$z - t = \text{constant}. \tag{137}$$

The characteristic lines or curves relating to a partial differential equation may be considered as the locus along which initial discontinuities in the derivatives of the entities involved will propagate in our case, in the (z, t) plane. For example, for the equation describing the equation of the wavefront, the rays will constitute the characteristics of the system. The reader is urged to look into references [15] and [16] for more information on the subject. Hence, moving along these characteristics we integrate the set (123) and (124) and obtain the coupled integral equations

$$\varphi_1(z, t) = \varphi_1^0(z + t) + \int_0^t \frac{k'}{2k}(z+t-s)\varphi_2(z+t-s, s)\, ds \quad (138)$$

and

$$\varphi_2(z, t) = \varphi_2^0(z - t) - \int_0^t \frac{k'}{2k}(z-t+s)\varphi_1(z-t+s, s)\, ds \,. \quad (139)$$

These two satisfy the differential equations and the initial conditions.

It has been demonstrated in Reference [4] that this coupled set is uniquely solvable and that the solutions can be obtained as the Neumann iterative series. The series is shown to converge for certain conditions on $k'/2k$ such as

$$\int_{-\infty}^{\infty} \left(\frac{k'}{2k}\right) dz < \text{a constant, say } 1 \text{ or } \frac{\pi}{2} \,. \quad (140)$$

If, however, we use the Laplace transform technique, we convert Equation (108), by taking the L-transform with respect to the variable t, into a time independent equation with a source term. We have

$$\frac{\partial^2\Psi}{\partial x^2} - k^2 s^2 \Psi(x, s) = -k^2(x)[\Psi_t(x, 0) + s\Psi(x, 0)] \,. \quad (141)$$

This type of equation with the source term has been treated earlier in Chapter V. We know the right-hand side of (141) as a function of x given by the initial conditions (112) and (113).

It is easily seen that from Equations (123) and (124) we can write the equation corresponding to n-th order solutions and from the coupled integral Equations (138) and

(139) we can obtain equations similar to the Bremmer series type solutions.

7. WIENER INTEGRALS

We refer to the work of Kae for a historical account of the Wiener integral, see References [17] and [6] for introductions to the mathematical theory and its physical applications. Methods of measure theory and integration in function spaces are widely applied in theories of heat diffusion quantum mechanical computations, and of random processes. Let x(t) denote a Brownian motion process along the x-axis starting from x_0 at time $t = 0$, subject to white noise impulses. The probabilities of its position at various times are governed by (ignoring the inertia of the particles) the diffusion equation

$$\frac{\partial\pi(x, t)}{\partial t} = D \frac{\partial^2\pi(x, t)}{\partial x^2}, \tag{142}$$

where D, the diffusion coefficient, is given by the well-known Einstein relations. The probabilities for occupying various positions are given by

$$\pi(x, t|x_o\ 0) = \frac{1}{(4\pi Dt)^{\frac{1}{2}}} \exp - \frac{(x - x_0)^2}{4Dt}, \tag{143}$$

is called the transition probability density. The probabilities of finding the particle at various times t, $0 < t_1 < t_2 < \cdots$ in the region between α_1 and β_1 at time t_1, in the region between α_2 and β_2 at time t_2, etc., are given by

$$\int_{\alpha_1}^{\beta_1} \cdots \int_{\alpha_n}^{\beta_n} \pi(x_1\ t_1 \mid x_0)\pi(x_2, t_2-t_1 \mid x_1)\cdots\pi(x_n, t_n - t_{n-1} \mid x_{n-1})\, dx_1 \cdots dx_n . \tag{144}$$

The above can be used to construct a probability measure on the sample space Ω of all real valued function $x(t)$; $0 < t < \infty$ which satisfy $x(0) = x_0$. We take the quantity (144) as the measure of the 'elementary set' of function $x(t) \in \Omega$ which satisfy the conditions

$$\alpha_1 < x(t) < \beta_1; \quad \alpha_2 < x_Z < \beta_2 \cdots \qquad \alpha_n < x(t) < \beta_n \,. \tag{145}$$

It is easily seen that the measure (144) assigned to the set (145) of quasifunctionals satisfy the usual consistency conditions. Also a Borel field can be generated by these elementary sets and their compliments. Extending these ideas suitably the Weiner measure concentrated on the subspace of continuous functions $x \in \Omega$ is generated. Employing this linear measure one can define the Wiener integrals. Using the concept of Wiener integrals one can solve the partial differential equations of the type

$$\frac{\partial\Psi(x, t)}{\partial t} + V(x)\Psi = \frac{\partial^2\Psi}{\partial x^2} \quad \text{with} \quad \Psi(x, 0) = \delta(x-x_0) \,. \tag{146}$$

These are also related to the path integral formalism of Feynman widely used in quantum mechanical computation [8]. Equation (146) is the perturbed diffusion equation which is met with in many physical and engineering problems. The path integral formalism is widely employed to derive the general features of the solution. It is also extremely useful when $V(x)$ is random, and we are interested in the moments of the solution $\Psi(x, t)$. The reader is referred to the literature cited in references [5], [6], [8] and [17] - [20]. These ideas will be presented in a subsequent volume.

REFERENCES

[1] Bellman, R., R. Kalaba, and G.M. Wing, 'Invariant Imbedding and Neutron Transport Theory, IV - Diffusion as a Limiting Case', Journal of Mathematics and Mechanics 9 (1960), 933.

[2] Bellman, R., R. Kalaba, and J. Lockett, Numerical Inversion of Laplace Transform, American Elsevier Publ. Co., Inc., New York, 1966.

[3] Scott, M.R., Invariant Imbedding and Its Applications to Ordinary Differential Equations, Addison-Wesley Publ. Co., Reading, Mass., 1973.

[4] Verheggen, Theo, Doctoral Thesis, Series Solutions of The Wave Equations Based on WKB Approximation, Netherlands (1974).
[5] Gelfand, I.M. and A.M. Yaglom, 'Integration in Functional Spaces and Its Applications in Quantum Physics', Journal of Mathematical Physics 1 (1960), 48.
[6] Kac, M., Probability and Related Topics in Physical Sciences, Wiley Interscience Division, John Wiley & Sons, Inc., New York, 1959.
[7] Srinivasan, S.K. and R. Vasudevan, Introduction to Random Differential Equations and Their Applications, American Elsevier Publ. Co., Inc., New York, 1971.
[8] Feynman, R.P. and A.R. Hibbs, Quantum Mechanics and Path Integrals, McGraw-Hill, New York.
[9] Ramakrishnan, A., R. Vasudevan, and P. Rajagopal, 'Ambigeneous Stochastic Process', Journal of Mathematical Analysis and Applications 1 (1960), 145.
[10] Weinberg, A.M. and E.P. Wigner, The Physical Theory of Neutron Chain Reactor, University of Chicago Press, Illinois, 1958.
[11] Chandrasekhar, S., 'Stochastic Problems in Physics and Astronomy', Rev. Mod. Phys. 15 (1945), 1.
[12] Bateman, H., in A. Erdelyi (ed.), Tables of Integral Transforms, McGraw-Hill, New York, 1952.
[13] Bellman, R. and G.M. Wing, An Introduction to Invariant Imbedding, John Wiley & Sons, Inc., New York, 1975.
[14] Barley, P.B., 'Sturm-Liouville Eigenvalues, Via a Phase Function', SIAM Journal of Applied Mathematics 14 (1966), 242.
[15] Abott, M.B., The Method of Characteristics, American Elsevier Publ. Co., Inc., New York, 1966.
[16] Courant, R. and D. Hilbert, Introduction to Mathematical Physics, Vol. II, Chapter I, Interscience Publishers, New York, 1943.
[17] Kac, M., 'Wiener and Integration in Function Spaces', Bulletin of the American Mathematical Society 72 (1966), 52.
[18] Brush, S.G., 'Functional Integral and Statistical Physics', Reviews of Modern Physics 33 (1961), 79.
[19] Arthurs, A.M., in Functional Integration and Its Applications, Clarendon Press, Oxford, 1975.
[20] Dashen, R., 'Path Integrals for Waves in Random Media', Jounal of Physics 20 (1979), 898.

CHAPTER VIII

ASYMPTOTIC PROPERTIES

1. INTRODUCTION

In this chapter we will be concerned with the asymptotic behavior of the solutions of the Schrödinger equation. This means we are interested in the modifications introduced into the wave function when it goes through a potential region. Instead of computing the solution in the entire region and determining its asymptotic form to find the phase shift introduced by the effect of the potential, we can write down an equation governing the phase shift itself. This turns out to be a nonlinear equation with an initial condition. This need not surprise us since these phase shifts are related to the reflection functions which are necessary to describe the spherical partial waves of the radial Schrödinger equation collapsing inwards and returning reflected from the scattering center. Reflection functions as we have seen already satisfy nonlinear equations.

The phase method is a particular way of relating the solution of the equation without the potential to the solution with the potential in the asymptotic region where the potential had died off. This can be done by computing the phase shift as a function of the extension of the interaction region. Hence, an analysis by imbedding in the length of the potential region is called for. Once the phase function or the effect of the potential is known as a function of the distance, it can be integrated to infinity to yield the phase shift.

The phase equation was introduced by Morse and Allis [1]. Also later different types of equations for the phase function were developed by Levy and Keller [2], Franchetti and Dashen [3]. Calogero's contribution to this field added new results to potential scattering [4]. In the first section we describe the method due to Wing [5] to arrive at a phase equation by a direct imbedding procedure. We then go on in the next section to relate the phase equation and the equations for the reflection functions for an incident wave going into a potential region, and getting reflected back.

If the wave number possesses smooth joining conditions at the point where potential is switched off, we obtain an imbedding equation, quite different from what we arrive at by an abrupt cut off of the potential. In the third section, we start with the Lippman-Schwinger integral equation for the scattering solution and show that by an imbedding of the Jost [6] function we can obtain the familiar nonlinear differential equation for the phase function.

2. ASYMPTOTIC BEHAVIOR OF THE SOLUTIONS OF THE SCHRÖDINGER EQUATION

The asymptotic behavior of the solutions of the Schrödinger equation becomes important where we are interested in the phase shifts introduced into the wave on passing through a potential region or region of interaction. We will be interested here with the simple problem of scattering in a scalar potential and hence discuss methods of computing it for a one-dimensional problem. Imbedding equations for the phase function can be derived in different ways. We will first describe the method due to Wing [5]. We will also describe other methods to obtain [2, 3, 4] the phase equations.

Let us study the equation

$$L(u) = \frac{d^2u}{dt^2} + u = f(t)u \tag{1}$$

with initial values

$$u(0) = d_1 \tag{2}$$

$$u'(0) = d_2. \tag{3}$$

The function f(t) is assumed continuous and bounded on the entire interval $t \geq 0$, where the conditions

$$\int_0^\infty |f(t)|dt < \infty \tag{4}$$

hold. The complete solution of Equation (1) can be written as was shown in Chapter II, for large t

$$u(t) = c_1 \sin t + c_2 \cos t + o(1) \tag{5}$$

$$= a \cos (t - \varphi) + o(1), \tag{6}$$

where a is called the amplitude and φ is the phase shift. One method of finding a, and φ is to integrate the equation far out to a large t, until it assumes the asymptotic form a cos $(t-\varphi)$ and compare the quantities. But to determine these quantities directly we can use the imbedding method by choosing carefully a family of related adjacent class of problems. We replace the initial conditions (2) and (3) by more convenient ones (choosing x as the initial position)

$$u(x) = \cos \theta \tag{7}$$

$$u'(x) = \sin \theta \tag{8}$$

and introduce the notation

$$u = u(t, x, \theta). \tag{9}$$

Thus the medium now extends from x to t, with initial values of u at x given by (7) and (8). If we choose to write the asymptotic form of u as

$$u(t, x, \theta) = a(x, \theta) \cos [t-x-\theta-\Psi(x, \theta)] + o(1), \tag{10}$$

where $\Psi(t, x, \theta)$ is called the phase shift function, we can relate Ψ for different values of x and θ.

The solution u of equation (1) or the equivalent integral equation for u is given by

$$u(t, x, \theta) = \cos (t-x-\theta) + \int_x^t \sin (t-y) f(y) u(y, x, \theta)\, dy. \tag{11}$$

This satisfies the initial conditions at t = x, and also the differential equation (1). This can be arrived at by a variation of parameters method [7] or otherwise. Classical differential equation theory assures us of the existence and continuity of $u_x(t, x, \theta)$ and $u_\theta(t, x, \theta)$ in t, x and θ.

Let us now restate the Gronwell-Bellman inequality [8] described in Chapter II and use that repeatedly to ensure convergence and existence of many quantities we meet in this treatment. We now state the Gronwell-Bellman inequality.

If $g(z)$ and $h(z)$ are continuous and nonnegative for $z > z_0$, for some $c > 0$, and if we have the relation

$$g(z) \leq c + \int_{z_1}^{z} g(z')h(z')dz' \qquad (12)$$

then

$$g(z) \leq c \exp\left\{\int_{z_1}^{z} h(z')\, dz'\right\}. \qquad (13)$$

As a straight application of this inequality we have from Equation (11)

$$|u(t, x, \theta)| \leq \exp \int_{x}^{t} |f(w)|dw \leq M, \qquad (14)$$

where M is a constant dependent only on f in view of Equation (4).

We now rewrite Equation (11) in a different form to obtain the asymptotic form more easily.

Equation (11) can be rewritten as

$$u(t, x, \theta) = \sin t \left\{\sin (x+\theta) + \int_{x}^{\infty} \cos w\, f(w)u(w, x, \theta)\, dw\right\} +$$

$$+ \cos t \left\{\cos (x+\theta) - \int_{x}^{\infty} \sin w\, f(w)u(w, x, \theta)\, dw\right\} -$$

$$- \int_{t}^{\infty} \sin (t-w)f(w)u(w, x, \theta) =$$

$$= b(x, \theta) \sin t + c(x, \theta) \cos t + \ell(t, x, \theta). \qquad (15)$$

We can now show that $b(x, \theta)$ and $c(x, \theta)$ have continuous partial derivatives with respect to x and θ, and also that

$$\lim_{t\to\infty} \ell(t, x, \theta) = 0 \qquad (16)$$

uniformly in x and θ.

First considering the boundedness of $\partial u/\partial x$ from Equation (11), we have

$$\frac{du}{dx} = \sin (t-x-\theta) - \sin (t-x)f(x)u(x, x, \theta) + \int_x^t \sin (t-w)f(w)u_x(w, x, \theta). \tag{17}$$

Since $u(x, x, \theta) = \cos \theta$, by the Gronwell-Bellman inequality, we have

$$\frac{\partial u}{\partial x} (t, x, \theta) \leq M_1 \tag{18}$$

where M_1 depends only on f and independent of t and θ. Now if we express $\partial b/\partial x$ as

$$\frac{\partial b}{\partial x} (x, \theta) = \cos (x+\theta) - \cos x \cos \theta f(x) + \int_x^\infty \cos w f(w)u_x(w, x, \theta) dw \tag{19}$$

we note that the continuity of $\partial b/\partial x$ follows from the above, u_x is bounded and the differentiation under the integral sign is justified. Similarly, the differentiability with respect to θ can be established. The quantity $c(x, \theta)$ can be similarly treated.

We can now write the solution $u(t, x; \theta)$ in the form

$$u(t, x, \theta) = A(x, \theta) \cos [t-x-\theta-\Psi(x, \theta)] + \ell(t, x, \theta). \tag{20}$$

Now Equation (15) can be reexpressed as

$$u(t, x, \theta) = \sqrt{b^2(x, \theta) + c^2(x, \theta)}\left\{\frac{b(x, \theta)}{\sqrt{b^2+c^2}} \sin t + \frac{c(x, \theta)}{\sqrt{b^2+c^2}} \cos t\right\} + \ell(t, x, \theta) = A(x, \theta)\cos [t-x-\theta-\psi(x, \theta)] + \ell(t, x, \theta). \tag{21}$$

Since b and c cannot be simultaneously zero,

$$A > 0. \tag{22}$$

From the definitions of b and c in Equations (15) we know that $b(x, \theta) = \sin(x+\theta)$ and $c = \cos(x+\theta)$ in the limit that $x \to \infty$. Hence,

$$\lim_{x\to\infty} A = \sqrt{c^2 + b^2} = 1. \tag{23}$$

Moreover, we recall that we keep t always $\geq$ x and hence when x becomes large Equation (15) yields

$$u(t, x, \theta) = \cos(t-x-\theta) \tag{24}$$

since

$$\left.\ell(t, x, \theta)\right\}_{x\to\infty} = 0$$

and

$$A(x, \theta)_{x\to\infty} = 1. \tag{26}$$

Thus we require that

$$\Psi(x, \theta)_{x\to\infty} = 0. \tag{27}$$

This can be seen if we rewrite Equation (21) as

$$A(x, \theta)\cos(t-\bar{\Psi}) + \ell(t, x, \theta), \tag{28}$$

where

$$\sin\bar{\Psi} = \frac{b(x, \theta)}{A(x, \theta)}; \quad \bar{\Psi} = \theta + x + \Psi. \tag{29}$$

Hence

$$\sin \Psi = \sin (\bar{\Psi} - \overline{\theta + x})$$

$$= \sin \bar{\Psi} \cos (\theta + x) - \cos \bar{\Psi} \sin (\theta + x) \qquad (30)$$

$$= \frac{b(x, \theta)}{A(x, \theta)} \cos (\theta + x) - \frac{c(x, \theta)}{A(x, \theta)} \sin (\theta + x). \qquad (31)$$

In the limit that $x \to \infty$,

$$A \sim 1$$

$$b \sim \sin (x + \theta) \qquad (32)$$

$$c \sim \cos (\theta + x).$$

Hence $\sin \Psi$ tends to zero. Hence $\Psi = m\pi$, where m can be any integer. If we choose that branch of Ψ for which $m = 0$, we have $\Psi(x, \theta) = 0$ as x tends to large values. It can be assumed that $\Psi(x, \theta)$ is also smooth and continuously differentiable in its variables.

If we had not started with the differential Equation (1), but instead with the equation

$$\frac{d^2u}{dt^2} + k^2u = f(t)u \qquad (33)$$

we would have obtained instead of (15) solution of the form

$$u(t, x, \theta) = b(x, \theta)u_1(t) + c(x, \theta)u_2(t) + \ell_1(t, x, \theta), \qquad (34)$$

where u_1 and u_2 are the two independent solutions of the equation

$$\frac{d^2u}{dt^2} + k^2u = 0, \qquad (35)$$

such that the Wronskian

$$W(u_1, u_2) = 1. \qquad (36)$$

Then b_1, c and ℓ will involve integrals over these functions u_1 and u_2 instead sin t and cos t functions.

To find the partial differential equations for $A(x, \theta)$ and $\Psi(x, \theta)$, we compute first the following quantities from Equation (11) and making use of Taylor expansion up to first order in Δ.

$$\begin{aligned} u(x+\Delta, x, \theta) &= u(x, x, \theta) + u'(x, x, \theta)\Delta \\ &= \cos\theta + \Delta\sin\theta + O(\Delta), \end{aligned} \tag{37}$$

$$u'(x+\Delta, x, \theta) = \sin\theta + \Delta\{f(x) - 1\}\cos\theta + O(\Delta). \tag{38}$$

To carry out the imbedding analysis we want an adjacent problem. Let us start with $u(x+\Delta, x+\Delta, \theta') = \cos\theta'$ and $u'(x+\Delta, x+\Delta, \theta') = \sin\theta'$ and relate them to that obtained starting from x, by the following relations

$$u(x+\Delta, x, \theta) = \eta(x, \theta)u(x+\Delta, x+\Delta, \theta') = \eta(x, \theta)\cos\theta', \tag{39}$$

$$u'(x+\Delta, x, \theta) = \eta(x, \theta)u'(x+\Delta, x+\Delta, \theta') = \eta(x, \theta)\sin\theta'. \tag{40}$$

These also define the quantity $\eta(x, \theta)$.

Now the quantities η and θ' can easily be obtained from Equations (39) and (40).

$$\frac{[u(x+\Delta, x, \theta)]^2 + [u'(x+\Delta, x, \theta)]^2}{\eta^2(x, \theta)} = 1. \tag{41}$$

Hence, using Equations (37) and (38)

$$\eta(\theta, x) = 1 + \cos\theta\sin\theta\, f(x)\Delta + O(\Delta). \tag{42}$$

Also

$$\sin\theta' = \frac{\sin\theta + \Delta[f(x) - 1]\cos\theta}{\eta(x, \theta)}. \tag{43}$$

Therefore, up to the first order in Δ we have

$$\theta' = \theta + (f(x)\cos^2\theta - 1)\Delta + O(\Delta). \tag{44}$$

By the linearity of the original initial value problem

$$u(t, x, \theta) = \eta(x, \theta)u(t, x+\Delta, \theta'). \tag{45}$$

It is now a very simple matter to substitute the solutions $u(t, x, \theta)$ and $u(t, x + \Delta, \theta')$ with different initial values for x and θ'. In view of Equation (20), we can write

$$A(x, \theta) \cos [t-x-\theta-\Psi(x,\theta)] + \ell(t, x, \theta)$$

$$= \eta(x, \theta)\{A(x+\Delta, \theta') \cos [t-x-\Delta-\theta'-\Psi(x+\Delta, \theta')]\} + \\ + \ell(t, x+\Delta, \theta') \tag{46}$$

Therefore, we have

$$-\theta-\Psi(x, \theta) = -\Delta-\theta'-\Psi(x+\Delta, \theta') \tag{47}$$

$$A(x, \theta) = \eta(x, \theta)A(x+\Delta, \theta'). \tag{48}$$

We have already seen that the various partial derivates of A and Ψ exist and are continuous. We now expand $\psi(x+\Delta, \theta')$ and $A(x+\Delta, \theta')$ around the arguments x and θ and utilize Equations (42) and (44), and we arrive at

$$\frac{\partial\Psi}{\partial x} + \frac{\partial\Psi}{\partial\theta} [f(x) \cos^2\theta - 1] = -f(x) \cos^2\theta \tag{49}$$

$$\frac{\partial A}{\partial x} + \frac{\partial A}{\partial\theta} [f(x) \cos^2\theta - 1] = \cos\theta \sin\theta f(x)A(x, \theta). \tag{50}$$

We also know that

$$\Psi(x, \theta)_{x\to\infty} = 0 \text{ uniformly in } \theta \tag{51}$$

and

$$A(x, \theta)_{x\to\infty} = 1 \text{ uniformly in } \theta \tag{52}$$

The system of partial differential equations can be handled by the method of characteristics. From (49) and (50) we have

$$\frac{dx}{ds} = 1; \frac{d\theta}{ds} = (f \cos^2\theta - 1); \frac{d\Psi}{ds} = -f(x) \cos^2\theta. \tag{53}$$

Hence choosing $s = x$, and $s = s_0$ when $x = x_0$

$$\Psi(x_0, \theta_0) = \int_{x_0}^{\infty} f(r) \cos^2 \theta(r)\, dr \tag{54}$$

and $\theta(s)$ is obtained from the differential equation

$$\frac{d\partial}{ds} = f(s) \cos^2 \theta(s) - 1 \tag{55}$$

with $\theta = \theta_0$ at $s = s_0$. Similarly the integral for $A(x_0, \theta_0)$ is given by

$$A(x_0, \theta_0) = \exp\left\{\frac{1}{2}\int_{x_0}^{\infty} f(r) \sin(2\,\theta(r))\,dr\right\} \tag{56}$$

This set of equations can be computed numerically in most cases and the solution is obtained for large t values.

$$u(t, x_0, \theta_0) = A(x_0, \theta_0) \cos\left[t - x_0 - \theta_0 - \Psi(x_0, \theta_0)\right] \tag{57}$$

can be arrived at for large t. The imbedding analysis for operators more complicated than that in Equation (1) was carried out in reference [9] and for the Coulomb potentials in reference [10].

To make contact with the results obtained for equations governing the phase functions $\delta(s)$ arrived at by Levy and Keller [2], and Calogero [4] and others [3], we note from (53) that

$$\frac{d}{ds}[\theta(s) + \Psi(s)] = -1 \tag{58}$$

so that

$$\theta + \Psi = -s + c \tag{59}$$

$$\theta(s) + s + \Psi(s) = c. \tag{60}$$

If now we want the asymptotic expression (57) to be of the form

$$u(t, x_0, \theta_0) = \text{const} \times [\sin (t + \delta(\infty)] \tag{61}$$

we set

$$\theta(s) + s = \frac{\pi}{s} - \delta(s). \tag{62}$$

Hence, we have

$$\frac{\pi}{2} - \delta(s) + \Psi(s) = c. \tag{63}$$

If s approaches ∞ we know $\Psi(s) \to 0$ (Equation (51) and $\delta(s)$ becomes $\delta(\infty)$ and

$$c = \frac{\pi}{2} - \delta(\infty) \text{ and } s + \theta(s) + \Psi(s) = \frac{\pi}{2} - \delta(\infty). \tag{64}$$

Substituting for $x + \theta(x) + \Psi(x, \theta)$ in (57), we have

$$u(t) = \text{const} \times \sin[t + \delta(\infty)]. \tag{65}$$

Hence, the $\theta(s)$ in our Equation (62) should be set equal to

$$\theta(s) = -s + \frac{\pi}{2} - \delta(s). \tag{66}$$

Therefore substituting in Equation (55), we obtain the nonlinear differential equation for $\delta(s)$ the phase function

$$\frac{d\delta}{ds}(s) = -f(s) \sin^2 [s + \delta(s)]. \tag{67}$$

If we set $\theta(0) = \pi/2$, then from (66) the initial condition $\delta(0) = 0$ is automatically obtained. Also we have

$$\Psi(0, \pi/2) = -\delta(\infty) \tag{68}$$

3. THE PHASE APPROACH

The determination of the phase shift when a wave passes through a potential region is important in quantum mechanical problems. Hence, to arrive at the differential equation for the phase function a simpler approach has been adopted by Calogero [4]. We start with the S-wave Schrödinger (reduced) equation in one dimension.

$$u''(r) + k^2 u(r) = v(r)u(r), \tag{69}$$

k being a constant. All the necessary conditions on $v(r)$ are assumed so that v is a 'regular' potential presenting no problems. The solution u goes as r when we approach zero and in the asymptotic region,

$$u(r) \to \text{constant} \times [\sin (kr + \delta)] \tag{70}$$

is called the phase shift. The scattering phase shifts vanish when the potentials vanish and the knowledge of phase shifts is equivalent to knowing all about the scattering amplitudes in a region of interaction. The solution of (69) can be written as the following integral equation with the proper behavior at zero and at ∞

$$u(r) = \sin kr + \frac{1}{k} \int_0^r \sin k(r - s) v(s) u(s)\, ds. \tag{71}$$

Let us introduce two other functions, C_0 and S_0, defined by

$$S_0(r) = -\frac{1}{k} \int_0^r dr'\, v(r') \sin kr'\, u(r') \tag{72}$$

$$C_0(r) = 1 + \frac{1}{k} \int_0^r dr'\, v(r') \cos kr'\, u(r'). \tag{73}$$

From (71), (72), and (73) we can write down the solution $u(r)$ as

$$u(r) = C_0(r) \sin kr + S_0(r) \cos kr. \tag{74}$$

As r tends to ∞,

$$u(r) \to C_0(\infty) \sin kr + S_0\, (\infty) \cos kr \tag{75}$$

$$= \text{const} \times [\sin (kr + \delta_0)], \tag{76}$$

where

$$\tan \delta_0 = \frac{S_0(\infty)}{C_0(\infty)}\,. \tag{77}$$

Let us call $t(r) = S_0(r)/C_0(r) = \tan \delta(r)$ such that

$$t(r)_{r \to \infty} = \tan \delta_0. \tag{78}$$

If we differentiate (72) and (73) and substitute for u(r) from (74) on the right-hand side, we obtain

$$S_0' = -k^{-1}v(r)\left[C_0(r)\sin^2 kr + S_0 \sin kr \cos kr\right] \tag{79}$$

$$C_0' = k^{-1}v(r)\left[C_0 \sin kr \cos kr + S_0 \cos^2 kr\right]. \tag{80}$$

Multiplying (79) by C_0 and (80) by S_0, subtracting (80) from (79) and dividing by C_0^2, we obtain the differential coefficient of t(r). After some algebra, we obtain

$$\delta'(r) = -k^{-1}v(r)\,[\sin^2(kr + \delta(r))]. \tag{81}$$

$\delta(r)$ is called the phase function while δ_0 obtained as the limit of $\delta(r)$ as r goes to ∞ is called the phase shift. From the definitions of $C_0(r)$ and $S_0(r)$ we see that $S_0(r)$ goes to zero and $C_0(r)$ goes to 1 at r = 0. Hence the initial condition for (81) is

$$\delta(0) = 0. \tag{82}$$

Thus, if we integrate up to $r = \infty$ the nonlinear first order Equation (81) with the initial condition (82) we obtain the phase shift δ_0 corresponding to the potential region specified by v(r), and characterized by the parameter k, related to the energy of the Schrödinger particle. Similar equations can be derived for higher partial wave equations which complicate the potential function. Even equations describing nonlocal interactions can be handled. The reader can look up reference [4] for further information.

The imbedding approach to compute the phase shifts can also be done in another fashion as we have described in earlier chapters. For the radial Schrödinger equation, with a spherically symmetric potential, the incident plane wave can be expressed as a sum of partial spherical waves individually collapsing onto the center of force and being reflected each with its reflection amplitudes of modified phases. The complete set of reflection amplitudes then describe the complete scattering phenomenon. The reader is referred to the book by Schiff [11] for understanding the splitting into partial waves and arriving at the 'radial'

Schrödinger equation which is a scalar equation in one variable 'r'. The reduced equation will be of the form

$$u''_\ell + k_\ell^2 u_\ell = 0, \tag{83}$$

where

$$k_\ell^2 = E - v(r) - \frac{\ell(\ell+1)}{r^2}, \tag{84}$$

and $E = k_1^2$ a (constant), $v(r)$ representing the potential in the region and $[\ell(\ell+1]/r^2$ is the centrifugal barrier for the ℓ-th partial wave. The solutions u_ℓ have the property that

$$u_\ell(0) = 0 \tag{85}$$

$$u_\ell(r) \underset{r\to\infty}{\to} e^{-ik_1 r} + R_\ell e^{ik_1 r}. \tag{86}$$

In fact, if we construct the solution of the Schrödinger equation by a linear combination of the Jost solutions, the solution at infinity, when the potential has died off, is given by [12].

$$u_\ell(k, r) \underset{r\to\infty}{\simeq} \left(\frac{F_\ell(k)}{F_\ell(-k)} e^{ikr} - e^{-ikr}\right), \tag{87}$$

where $F_\ell^*(k) = F_\ell(-k)$ and they are called Jost functions [13] and the phase shift δ is given by

$$e^{2i\delta} = (-1)^\ell \frac{F_\ell(k)}{F_\ell(-k)}. \tag{88}$$

We will not go into the details of these Jost functions and Jost solutions but just point out that R, the reflection function in our Equation (86), is to be related to these phase shifts. If we solve Equation (83) with v being completely absent, we can easily deduce that the centrifugal potential produces at large distances a reflection amplitude $e^{-(\ell+1)\pi i}$ equal to +1 or -1, depending on the parity of the partial wave. Hence, for the S wave, i.e., for $\ell = 0$, the reflection amplitude is equal to −1, corresponding to a perfect reflector at the origin. The effect of the potential is to modify the reflection, which may then be written as

$$R_\ell = e^{-i\pi(\ell+1)+2i\delta_\ell} . \tag{89}$$

If we refer to our expression for $\rho(x)$ derived in Equation (17) of Chapter IV, it is the reflection amplitude for unit input amplitude at $r = x$ and $\Psi(x) = 1 + \rho(x)$ at $r = x$. Here we have an input amplitude $e^{-ik_1(r-x)}$ which comes out as $R\, e^{+ik_1(r-x)}$. Hence, taking into account the phase modification due to the potential, R is related to ρ as

$$R_\ell(r) = e^{-2ik_1 r} \rho_\ell(r) = e^{2i[\delta_\ell(r)-(\ell+1)\frac{\pi}{2}]} . \tag{90}$$

Therefore

$$\rho_\ell(r) = e^{2i[k_1 r-(\ell+1)\frac{\pi}{2} + \delta_\ell(r)]} . \tag{91}$$

$\delta_\ell(r)$ is the phase function which becomes the phase shift as $r \to \infty$. We know from Chapter IV that $\rho(r)$ satisfies the differential equation

$$\rho'(r) = \frac{K'_\ell}{2K_\ell}\left(1-\rho_\ell^2\right) + 2iK_\ell\rho, \tag{92}$$

where

$$K_\ell^2 = \left[k_1^2 - v(r) - \frac{\ell(\ell+1)}{r^2}\right] . \tag{93}$$

Let us now consider the S wave scattering i.e., the case when $\ell = 0$.

$$\rho'_0(r) = \frac{K'_0}{2K_0}(1-\rho_0^2) + 2iK_0\rho \tag{94}$$

and since $R_0(0)$ has been argued to be equal to −1, the initial condition for ρ_0 corresponding to a perfect reflector at the origin is

$$\rho_0(0) = -1. \tag{95}$$

Substituting the value of $\rho_0(r)$ in (91) into Equation (94) we obtain

$$\delta_0'(r) = -k_1 + K_0 + \frac{K_0'}{2K_0} \sin 2(k_1 r + \delta_0) \tag{96}$$

with

$$\delta_0(0) = 0. \tag{97}$$

Naming the variable $k_1 r + \delta(r)$ as $\alpha_0(r)$, we can write the differential equation for α as

$$\alpha_0'(r) = K_0 + \frac{K_0'}{2K_0} \sin 2\alpha. \tag{98}$$

However, when we have a classical turning point in the problem, i.e., if at a point r_0 the value of $k_0 = k_1^2 - v(r_0)$ becomes zero, there is a divergence in Equation (98). For the case in which there is a classical turning point, at r_0, $\alpha(r_0)$ can be shown to be zero for well-behaved potentials and α can be integrated from the point r_0. The reader is referred to the thesis by MacCallum [14].

If the potential is attractive there are no turning points. If the function K is well-behaved with no zeroes or singularities equation (94) for ρ_ℓ can be solved in an iterative fashion by the use of quasilinearization methods described earlier. However, we can also solve this Riccati equation by an order of scattering technique, elaborated in earlier chapters. Hence, we rewrite

$$\rho_{\ell,2N+1}'(r) = \frac{K_\ell'}{2K_\ell}\delta_{N,0} + 2iK_\ell\rho_{\ell,2N+1} - \frac{K_\ell'}{2K_\ell}\sum_{m=0}^{N-1}\rho_{\ell,2m+1}(r)\times$$

$$\times\,\rho_{\ell,2(N-m)-1} \tag{99}$$

with

$$\rho_{\ell,1}(0) = (-1)^{\ell+1}$$

$$\rho_{\ell,2N+1}(0) = 0 \quad \text{for} \quad N \neq 0$$

as initial conditions. The quantities $\rho_{\ell,2N+1}(r)$ are the reflected amplitudes coming out of the potential region having undergone exactly 2N+1 backscatterings inside the region. For example, with N = 0

$$\rho'_{\ell,1} = \frac{K'_\ell}{2K_\ell} + 2iK_\ell\rho_{\ell,1} \tag{100}$$

and for the value N = 1

$$\rho'_{\ell,3} = 2iK_\ell\rho_{\ell,3} - \frac{K'_\ell}{2K_\ell}\rho^2_{\ell,1} \; . \tag{101}$$

Similarly, for higher values of N. Hence, we see that the equations for all values of N form an infinite set of linear equations. In principle, one can solve them and find the total ρ_ℓ given by

$$\rho_\ell = \sum_{N=0}^{\infty} \rho_{\ell,2N+1}. \tag{102}$$

Hence for the S wave reflection, if we call

$$U_{2N+1}(r) = \rho_{0,2N+1}(r) = e^{2i(k_1 r+\delta_0-\frac{\pi}{2})} \tag{103}$$

the equation for N = 0

$$U'_1 = \frac{-K'_0}{2K_0} + 2iK_0U_1 \tag{104}$$

with

$$U_1(0) = -1. \tag{105}$$

The solution is easily obtained and we move on to the next order U_3, etc. If now we neglect $k'_0/2k_0$ as being very small the equation for log U_1 becomes

$$\frac{1}{2i}\frac{d}{dr} \log U_1 = K_0. \tag{106}$$

If we subtract $\bar{U}_1$ which pertains to the case when v the potential is equal to zero, we find the difference

$$\frac{1}{2i}\frac{d}{dr}\left[\log U_1 - \log \bar{U}_1\right] = (K_0 - k_1). \tag{107}$$

This quantity $1/2i\left[\log U_1 - \log \bar{U}_1\right]$ for large values of r becomes the W.K.B. phase shift for the potential V. Hence

$$\delta_{W.K.B.}(\infty) = \int_0^\infty (K - k_1)\, ds. \tag{108}$$

Once we find the quantity U_1 from (104), we substitute it in the equation for U_3

$$U_3' = 2iK_0U_3 - \frac{K_0'}{2K_0}\,U_1^2 \tag{109}$$

with initial condition $U_3(0) = 0$. Thus, all U_{2N+1} can be found in an iterative fashion. For details of obtaining the linearization of the Riccati equations the reader can look into reference [15] .

Equation (96) for $\delta(r)$ is quite different from the Equations (67) and (81) obtained for the phase function in earlier Sections 2 and 3 of this chapter. This is because the imbedding we have adopted in this section so far is of a different nature from that of Section 2 and 3. We have imposed in deriving the equation for ρ in Chapter IV the restriction concering the smooth joining of the media at $r = x$. The imbedding of Sections 2 and 3 relate to the process in which we cut off the potential at $r = x$, abruptly. In earlier chapters we have treated the transport equations for u and v beams traversing the medium in opposite directions. Let us now consider the wave equation

$$\Psi''(z) + k^2(z)\Psi = 0, \tag{110}$$

where k(z) the wave number in the region $z < x_0$ is k_0 and in the region $z > x$ is k_1. Let k(z) the refractive index in the region $x_0 < z < x$ be piecewise continuous and positive. We are not requiring the continuity of k at $z = x$ as we did in Chapter IV. Let a left moving wave impinge on the medium (from $z = +\infty$) at $z = x$ and the wave in the region $z > x$ is of the form

$$\Psi(z) = e^{-ik_1(z-x)} + \rho(x)\, e^{ik_1(z-x)} \tag{111}$$

and for $z < x_0$

$$\Psi(z) = \tau(x)\, e^{-ik_0(z-x_0)} . \tag{112}$$

Equations (111) and (112) represent the situation when no other wave impinges on the medium from $z < x_0$ at $z = x_0$. However, if there is a reflection at $z = x_0$, this condition is different. $\rho(x)$ and $\tau(x)$ as we have seen already are the reflection and transmission functions. $\rho(x)$ and $\tau(x)$ might not satisfy the same type of equations as we found in Chapter IV since we have not imposed the differentiability conditions of $k(z)$ as was done in Chapter IV. Let us now introduce two quantities, u which suggests a wave moving, to the right, and v, a wave moving to the left. Let us define them as

$$u = \frac{1}{2ik_0}\left[\Psi'(z) + ik_0\Psi(z)\right] \tag{113}$$

$$v = -\frac{1}{2ik_1}\left[\Psi'(z) - ik_1\Psi(z)\right]. \tag{114}$$

If there is a wave with wavenumber $k_0 (z < x_0)$, a rightmoving wave has the form $e^{ik_0 z}$; in that region $u = e^{ik_0 z}$ as seen from Equation (113). A similar remark holds for v. Let us now obtain the transport equations for these two oppositely moving beams. The wave equation can be written as a vector equation for a vector with Ψ and Ψ' as components.

$$\frac{d}{dz}\vec{\Psi} = \frac{d}{dz}\begin{pmatrix}\Psi \\ \Psi'\end{pmatrix} = \begin{pmatrix}0 & 1 \\ -k^2 & 0\end{pmatrix}\begin{pmatrix}\Psi \\ \Psi'\end{pmatrix}. \tag{115}$$

Making a transformation of the vector $\vec{\Psi}$ to a vector $\vec{\varphi}$ with components u and v, we arrive at the coupled differential equations for u and v

$$\vec{\varphi} = p\vec{\Psi};\ p = \begin{bmatrix} \frac{1}{2} & \frac{1}{2ik_0} \\ \frac{1}{2} & \frac{1-}{2ik_{10}} \end{bmatrix} \tag{116}$$

$$\frac{d}{dz}\begin{pmatrix} u \\ v \end{pmatrix} = p\,\frac{d}{dz}\begin{pmatrix} \Psi \\ \Psi' \end{pmatrix} = p\begin{bmatrix} 0 & 1 \\ -k^2 & 0 \end{bmatrix} p^{-1}p \begin{bmatrix} \Psi \\ \Psi' \end{bmatrix} \tag{117}$$

Hence,

$$u' = \frac{i}{k_0(k_0+k_1)}\left\{k_0\left[k^2 + k_0k_1\right]u + k_1\left[k^2 - k_0^2\right]v\right\} \tag{118}$$

$$-v' = \frac{i}{k_1(k_0+k_1)}\left\{k_0\left[k^2 - k_1^2\right]u + k_1\left[k^2 + k_0k_1\right]v\right\} \tag{119}$$

For the incident wave $e^{-ik_1(z-x)}$ at $z = x$, at the interface x,

$$\begin{aligned} \Psi(x) &= 1 + \rho(x) \\ \Psi'(x) &= -ik_1[1 - \rho(x)] \end{aligned} \tag{120}$$

and at $z = x_0$

$$\begin{aligned} \Psi(x_0) &= \tau(x) \\ \Psi'(x_0) &= ik_0\tau(x). \end{aligned} \tag{121}$$

Hence for this problem the boundary conditions are

$$u(x_0) = 0; \quad v(x) = 1. \tag{122}$$

If we now relate u and v by

$$u(z) = r(z)v(z)\ , \tag{123}$$

we can as explained in earlier chapters arrive at the equation for $r(x)$ making use of the coupled set (118) and (119)

$$r'(x) = \frac{i}{(k_0+k_1)}\left\{\frac{k_1}{k_0}\left[k^2 - k_0^2\right] + 2\left[k^2 + k_0k_1\right]r(x) + \right.$$

$$\left. + \frac{k_0}{k_1}\left[k^2 - k_1^2\right]r^2(x)\right\}, \tag{124}$$

and $r(0) = 0$ in view of Equations (122) and (123).

However, it must be noted that $r(x)$ gives $u(x)$ at $z = x$, for unit input, and from Equation (113) and (120) we have

$$r(x) = \frac{(k_0 + k_1)\rho(x) + (k_0 - k_1)}{2k_0} . \tag{125}$$

In other words

$$\rho(x) = \frac{2k_0}{k_1+k_0}\, r(x) + \frac{k_1 - k_0}{k_1 + k_0} . \tag{126}$$

From the above relation between $\rho(x)$ and $r(x)$, we easily see that $\rho(x)$ satisfies the differential equation

$$\rho'(x) = \frac{i}{2k_1}\left\{(k^2(x) - k_1^2) + 2\left[k^2(x) + k_1^2\right]\rho(x) + \right.$$

$$\left. + \left[k^2(x) - k_1^2\right]\rho^2(x)\right\} \text{ with} \tag{127}$$

$$\rho(x_0) = \frac{k_1 - k_0}{k_1 + k_0} .$$

If instead of the boundary condition (121) at $z = x_0$, we are given that $\Psi(x_0) = 0$ and $\Psi'(x_0) = 1$, $r(x_0)$ becomes $-k_{1/k_0}$.

When this value is substituted in Equation (126), we find that

$$\rho(x_0) = -1. \tag{128}$$

For the Schrödinger equation, we know that the quantity k represents

$$k^2(z) = k_1^2 - v(z) \tag{129}$$

in the case of S wave scattering. It is also well known that the wave function, which is the regular solution of the Schrödinger equation, satisfies the condition that $\Psi(0) = 0$ and $\Psi'(0) = 1$ at the origin. As we have explained earlier, the scattering in spherically symmetric potential can be imagined to be equivalent to the waves from large distance collapsing into the scattering center and getting reflected back with modified phases. Hence, for the incidence of a wave $e^{-ik_1 r}$ the reflection function $\rho(r)$ was found as

$$\rho(r) = e^{2i(k_1 r + \delta(r) - i\pi/2)}, \tag{130}$$

where $\delta(r)$ is the phase function for the S wave $\rho(0) = -1$, which is relevant to the initial conditions. Hence $\delta(0) = 0$, and substituting Equation (129) and (130) for the quantities $k(x)$ and $\rho(x)$ in Equation (127) we have the differential equation for the phase function $\delta(r)$ given by

$$\delta'(r) = -\frac{v}{k_1} \sin^2\left[k_1 r + \delta(r)\right] \tag{131}$$

with the initial condition corresponding to Equation (128)

$$\delta(0) = 0. \tag{132}$$

These were the equations obtained in Sections 1 and 2.

4. INTEGRAL EQUATION REPRESENTATION

Of course, there is another direct way of obtaining the imbedding equation for the phase shift by converting the Schrödinger equation into an integral equation. Without going into too many physical details, we will write down the integral equations corresponding to different initial and boundary conditions for the radial version of the Schrödinger equation for a radially symmetric potential for the ℓ-th partial wave

$$\left[-\frac{d^2}{dr^2} + \frac{\ell(\ell+1)}{r^2} + \gamma v(r)\right]\Psi_\ell(k, r) = k^2\Psi_\ell(k, r). \tag{133}^*$$

* $k^2 = E$, the constant energy given to the particle.

For the 'scattering' solution $\Psi_\ell(k, r)$, the corresponding integral equation can be obtained as (Lippman-Schwinger equation in a reduced form [16])

$$\Psi_\ell(k, r) = u_\ell(k, r) + \gamma \int_0^\infty ds\, G_\ell(k, r, r')v(r')\Psi_\ell(k, r')\, dr', \tag{134}$$

$G_\ell(k, r, r')$ is the Green's function for Equation (133) with suitable boundary conditions [12]. If we concentrate on the S wave 'scattering' solutions, we have

$$\Psi(k, r) = \sin kr + \varphi(k, r). \tag{135}$$

The solution Ψ goes to zero at $r = 0$. At $r = \infty$ there is no incoming wave due to the portion φ in Equation (135), i.e.,

$$\lim_{r\to\infty}\left[\varphi'(k, r) - ik\varphi(k, r)\right] = 0. \tag{136}$$

This means the Ψ has incoming contributions only due to u_ℓ. Also $\varphi(k, r)$ as r goes to zero becomes zero. These define the boundary values of the Green's function $G_\ell(k; r, r')$. For the case $\ell = 0$, i.e., for the S wave, $G(k; r, r')$ is given by

$$\begin{aligned} G(k, r; s) &= -e^{ikr}\frac{\sin ks}{k} \qquad r > s \\ &= -e^{iks}\frac{\sin kr}{k} \qquad r < s \end{aligned} \tag{137}$$

Thus, the scattering solution for the S wave Equation (135) is

$$\Psi(k, r) = \sin kr + \gamma\int_0^\infty G(k; r, s)v(s)\Psi(k, s)\, ds \tag{138}$$

$v(r)$ satisfies all the necessary conditions for the convergence of the solution. The potential dies off as we approach large distances.

The solution Ψ was r goes to infinity becomes from (137) and (138)

$$\Psi(r) = \frac{i}{2}\left(e^{-ikr} - Se^{ikr}\right), \tag{139}$$

where S then, represents the scattering effects from the potential which goes to unity in the absence of any potential interaction.

Let us now consider another solution Φ of the S wave equation called the regular solution with initial conditions

$$\Phi(k; r)\big|_{r=0} = 0 \tag{140}$$

$$\Phi'(k; r)\big|_{r=0} = 1. \tag{141}$$

To ensure these conditions, the Green's function is not the Green's function of equation (137) but is given by the following:

$$G^0(k, r, r') = \frac{\sin k\,(r-r')}{k} \quad r' < r$$
$$= 0 \quad r' > r \tag{142}$$

Hence, the integral equation for $\Phi(k; r)$ is

$$\Phi(k, r) = \frac{\sin kr}{k} + \frac{\gamma}{k}\int_0^r \sin k(r-r')v(r')\Phi(k, r')\,dr' \tag{143}$$

which is an integral equation of the Volterra type. Yet another solution of Equation (133) for $\ell = 0$, called the Jost solution $f(k; r)$ satisfying the condition at $r = \infty$ given by

$$\lim_{r\to\infty}\left[e^{-ikr}f(k, r)\right] = 1 \tag{144}$$

can be shown to satisfy the integral equation

$$f(k, r) = e^{ikr} + \gamma\int_r^\infty \frac{\sin k(r'-r)}{k}\, v(r')f(k, r')\,dr'. \tag{145}$$

Similarly $f(-k, r)$ is another solution of the S wave equation with the condition at infinity given by

$$\mathrm{Lim}\left[e^{-ikr} f(-k, r)\right]_{r=\infty} = 1. \tag{146}$$

It satisfies the same equation as (145) with the variable k replaced by -k. It is to be noted that f(+k; r) and f(-k; r) are two linearly independent solutions of the S-wave Schrödinger equation and any other solution like the regular solution $\Phi(k, r)$ can be expressed as a linear combination of f(+k, r) and f(-k, r). Hence,

$$\Phi(k, r) = af(k, r) + bf(-k, r). \tag{147}$$

The Wronskian of two functions f and g is defined as

$$W(f, g) = fg' - f'g. \tag{148}$$

If f and g are both solutions of the same second-order differential equation, then the Wronskian is a constant independent of r [7]. Hence, the Wronskian in such a case can be evaluated at any convenient value of r at which the functions f and g and their derivatives are known. We recall

$$\Phi'(0) = 1 \quad \text{and} \quad \phi(0) = 0. \tag{149}$$

Hence,

$$W\left[f(k, r), \Phi\right] = f(k) = 1 + \frac{\gamma}{k}\int_0^\infty \sin kr'\, v(r') f(k, r')\, dr' \tag{150}$$

$$W\left[f(-k, r)\ \Phi\right] = f(-k) = 1 + \frac{\gamma}{k}\int_0^\infty \sin kr'\, v(r') f(-k, r')\, dr', \tag{151}$$

where f(k) and f(-k) are called the Jost functions. Hence, taking the Wronskian of the left-hand side of Equation (147) with f(k, r) and f(-k, r), the coefficients a and b can be determined in terms of Jost functions and we have therefore

$$\Phi(k, r) = \frac{1}{2ik}\left[f(-k) f(k, r) - f(k) f(-k, r)\right]. \tag{152}$$

Therefore, as r goes to infinity, the solution $\Phi(k, r)$ becomes

$$\Phi(k, r) = \frac{f(k)}{2ik}\left[\frac{f(-k)}{f(k)}\, e^{ikr} - e^{-ikr}\right] =$$

$$= \frac{1}{2ik}\left[f(-k)e^{ikr} - f(k)e^{-ikr}\right]. \tag{153}$$

Let us now return to Equation (139) for the asymptotic value of the Ψ function, and note that quantity S which is a function of k, can be written down as

$$S = \exp\,(2i\delta). \tag{154}$$

This is because the S matrix is unitary. In other words, due to the conservation of flux (which has its origin in the hermiticity of the Hamiltonian), it happens that the magnitude of S is unity [11], hence, from Equation (138) (suitably defining S)

$$\Psi(k, r) = \frac{1}{2i}\left[Se^{ikr} - e^{-ikr}\right]. \tag{155}$$

Let us now consider the kernel in the integration of integral Equation (143) and rewrite the Green's function G(k, r, r') as

$$G^0(k, r, r') = G(k, r, r') + \frac{\sin kr}{k}\, e^{ikr'}. \tag{156}$$

Hence, the integral Equation (143) for Φ can be rewritten as

$$\Phi(k, r) = \frac{\sin kr}{k}\left[1 + \gamma\int_0^\infty e^{ikr'}v(r')\Phi(k, r')\,dr'\right]$$

$$+ \gamma\int_0^\infty G(k, r, r')v(r')\Phi(k, r')\,dr'. \tag{157}$$

In a similar fashion integral Equation (145) for f(k, r) can be rewritten as

$$f(k, r) = e^{ikr}\left[1 + \frac{\gamma}{k}\int_0^\infty \sin(ks)v(s)f(k, s)\, ds\right] +$$

$$+\gamma\int_0^\infty G(k, r, r')f(k, r')v(r')\, dr' \qquad (158)$$

$$= e^{ikr}f(k) + \gamma\int_0^\infty G(k, r, r')v(r')f(k, r')\, dr'.$$

Comparing Equation (138) with (158), they differ only in the forcing functions and we can easily write down the relation

$$f(k, r) = \frac{\Psi(k, r)e^{ikr}f(k)}{\sin kr}\,. \qquad (159)$$

Substituting this back into the equation for f(k), we arrive at the relation

$$f(k) = \left[1 - \gamma/k\int_0^\infty e^{iks}v(s)\Psi(k, s)\, ds.\right]^{-1}. \qquad (160)$$

Comparing Equations (157) and (158), we have the relation

$$\frac{f(k, r)\sin kr}{k}\left[1 + \gamma\int_0^\infty e^{ikr'}v(r')\Phi(k, r')\, dr'\right]$$

$$= \Phi(k, r)e^{ikr}f(k). \qquad (161)$$

Making use of Equation (150) and (161) we easily arrive at the relation

$$f(k) = 1 + \gamma\int_0^\infty e^{ikr'}\, v(r')\,\Phi(k, r')\, dr'. \qquad (162)$$

Therefore, we have from Equation (157)

$$\Phi(k, r) = \frac{\sin kr}{k}f(k) + \int_0^\infty \gamma G(k, r, r')\Phi(k, r)v(r')\, dr'. \qquad (163)$$

Comparing Equation (163) with (138), we see the relation between Φ and Ψ as

$$\psi(k, r) = \frac{k\Phi(k, r)}{f(k)} . \tag{164}$$

In view of Equations (153) and (155), which express the Ψ and Φ solutions at large distances after the potential has vanished, and taking into account relation (164), we conclude that

$$S = \frac{f(-k)}{f(k)} = e^{2i\delta}, \tag{165}$$

and since $f^*(k, r)$ and $f(-k, r)$ satisfy the same integral equation

$$f(-k) = f^*(k) = e^{i\delta} \; |f| . \tag{166}$$

This is known as the crossing symmetry relation. To establish this result for S(k) we need not have taken the trouble to find the expression for $\Phi(k, r)$ in terms of f(k, r) and f(-k, r) as in Equation (152). It is enough if we had noted that since $\Phi(k, r)$ is the solution of the differential equation involving only k^2 and the initial conditions $\Phi(k, 0)$ and $\Phi'(k, 0)$ do not depend on k, we have,

$$\Phi(k, r) = \Phi(-k, r). \tag{167}$$

The asymptotic value of Ψ contains S, which can be seen in view of Equations (138) and (155) to be given by the expression

$$S(k) = 1 - \frac{2i\gamma}{k} \int_0^\infty \sin kr' v(r') \Psi(k, r') \, dr' . \tag{168}$$

Because of relation (164), we replace $\Psi(k, r')$ and obtain

$$S(k) = 1 - \frac{2i\gamma}{k} \int_0^\infty \sin kr' v(r') \, \frac{k\Phi(k, r')}{f(k)} \, dr' \tag{169}$$

$$= \frac{f(k) - \gamma\int_0^\infty e^{ikr'}v(r')\Phi(k, r')\,dr' + \gamma\int_0^\infty e^{-ikr'}v(r')\Phi(k, r')\,dr'}{f(k)}$$

$$= \frac{1 + \gamma\int_0^\infty e^{-ikr'}v(r')\Phi(-k, r')\,dr'}{f(k)} = \frac{f(-k)}{f(k)}\,. \qquad (170)$$

Equation (138) is the Fredholm equation of the second kind for the solution $\Psi(k, r)$ and if $R(k, r, s, \gamma)$ is the resolvent of the kernel $G(k, r, s)v(s)$ in that equation, then the solution $\Psi(k, r)$ can be expressed as

$$\Psi(k, r) = \sin kr + \gamma\int_0^\infty R(k, r, s, \gamma)\sin(ks)ds. \qquad (171)$$

According to Fredholm theory the resolvent can be expressed as

$$R(k, r, s, \gamma) = \frac{N(k, r, s, \gamma)}{d(\gamma)}\,, \qquad (172)$$

where d is solely a function of γ and independent of r and s. This is called the Fredholm determinant corresponding to the given kernel. It is to be noted that $[k\Phi]$, in view of Equation (163), will satisfy an equation

$$[k\Phi(k, r)] = \sin[krf(k)] + \int_0^\infty R(k, r, s, \gamma)\sin(ks)ds\,f(k) \qquad (173)$$

which would be the same equation satisfied by Ψd if d, the Fredholm determinant, is equal to f(k) the Jost function. In fact this has to be so since $k\Phi(k, r)$ is identical to $\Psi(k, r)f(k)$ because of relation (164). It can be shown in more rigorous fashion by actually computing the Fredholm determinant for the given kernel.

In the above we have dealt with integral equations with the upper limit for r as infinite. The potential dies off and asymptotically the wave functions are free-wave functions with modified phases. We could imagine that the potential is cut off at a, and all the equations hold with the upper limit a for the variable r. Thus, we imbed the problem, in a, the extension of the potential region and since the phase shift δ is a function of k, and a, we arrive at an imbedding equation for the phase function $\delta(a)$. When we integrate this out to infinity, $\delta(\infty)$ becomes the phase shift for the given problem. Since $f(+k)$ is proportional to $e^{-i\delta(k,a)}$, we can start with Equation (160) for f(k) with the upper limit replaced by r = a

$$f(k, a) = \left[1 - \frac{\gamma}{k}\int_0^a e^{iks} v(s)\Psi(k, s, a)\, ds\right]^{-1}. \qquad (174)$$

Since $f(k, a) \simeq e^{-i\delta(k,a)}$, we find the imbedding equation for $\delta(k, a)$ by:

$$\frac{d}{da}\log f(k, a) = \frac{\frac{\gamma}{k} e^{ika} v(a)\Psi(k, a, a) + \frac{\gamma}{k}\int_0^a e^{iks} v(s)\Psi_a(k, s, a)\, ds}{\left[1 - \frac{\gamma}{k}\int_0^a e^{iks} v(s)\Psi(k, s, a)\, ds\right]} \qquad (175)$$

It is evident in this case that $\Psi(k, r, a)$ is a function of a, the upper limit, since it will satisfy integral Equation (138)

$$\Psi(k, r, a) = \sin(kr) + \gamma\int_0^a G(k, r, s)v(s)\Psi(k, s, a)\, ds. \qquad (176)$$

Hence, to obtain Ψ_a in Equation (175) which represents the variation of Ψ with respect to a, we find the revolvent $R(k, r, s, a)$ of the kernel in (176) and obtain Ψ_a as

$$\Psi_a(k, r, a) = \gamma R(k, r, a, a)\Psi(k, a, a). \qquad (177)$$

Let us compute $R(k, r, a, a)$ for the kernel $G(k, r, s)v(s)$

$$R(k, r, a, a) = G(k, r, a)v(a) + \gamma\int_0^a R(k, r, \ell, a)G(k, \ell, a)v(a)\, d\ell \tag{178}$$

$$= \frac{-\Psi(k, r\, a)e^{ika}}{k}\, v(a).$$

Combining the results in (178), (177) and (175), we obtain

$$\frac{d}{da}\log f(k, a) = \frac{\Psi}{k}\, e^{ika} v(a)\Psi(k, a, a). \tag{179}$$

From Equation (155) we know that Ψ assumes the value given by the following expression after exiting from the potential region,

$$\Psi(k, a, a) = e^{i\delta(k, a)} \sin\,[ka + \delta(k, a)]. \tag{180}$$

Hence, the variation of $\delta(k; a)$ with a is obtained by making use of (180) and (179) as

$$\delta'(a) = -\mathrm{Im}\,\frac{d}{da}\log f(k, a) = -\frac{\gamma}{k}\, v(a)\sin^2[ka + \delta(k, a)] \tag{181}$$

which is the same nonlinear equation obtained earlier for the phase function $\delta(k; r)$ in Equations (67) and (81). This can easily be applied to higher partial wave equations. Also, the concept of the phase shift has been extended to the case of non-local and momentum dependent potentials (see Reference [4]). The imbedding method can be applied to all these cases.

REFERENCES

[1] Morse, P.M. and W.P. Allis, Physical Review 44 (1933), 269.

[2] Levy, Bertram R. and Joseph B. Keller, 'Low-Energy Expansion of Scattering Phase Shifts for Long-range Potentials', Journal of Mathematical Physics 54 (1963).

[3] (a) Franchetti, S, Nuovo Cimento 6 (1957), 601.
(b) Dashen, Roger F., 'New Approach to Low-Energy Potential Scattering', Journal of Mathematical Physics 4 (1963), 388.

[4] Calogero, F., Variable Phase Approach to Potential Scattering, Academic Press, New York, 1967.

[5] (a) Wing, G.M. and R. Bellman, An Introduction to Invariant Imbedding, John Wiley and Sons, New York, 1975.

(b) Wing, G.M., 'Invariant Imbedding and the Asymptotic Behaviour of Solutions to Initial Value Problems', Journal of Mathematical Analysis and Application 9 (1964), 85.

[6] Newton, R.G., Journal of Mathematical Physics 1 (1960), 319.

[7] Coddington, Earl A., An Introduction to Ordinary Differential Equations, Prentice-Hall Inc., Englewood Cliffs, N.J., 1961.

[8] Beckenbach, Edwin F. and Richard Bellman, Inequalities, Springer Verlag, 1965.

[9] Hagin, F.G., 'Some Asymptotic Behaviour Results for Initial Value Problems: An Application of Invariant Imbedding', Journal of Mathematical Analysis and Application, 20 (1967), 540-564.

[10] Beyer, W.A., 'Asymptotic Phase and Amplitude for a Modified Coulomb Potential in Scattering Theory: An Application of Invariant Imbedding', Journal of Mathematical Analysis and Application 13 (1966), 348-360.

[11] Schiff, L.I., Quantum Mechanics, Third Edition, McGraw-Hill Book Company, 1968.

[12] Newton, R.G., Scattering Theory of Waves and Particles, McGraw Hill Book Co., 1966.

[13] (a) Jost, R. and A. Pais, 'On the Scattering of a Particle by a Static Potential', Physical Review 82 (1951) 840.

(b) von Res Jost, 'Bemerkungen zu der vorstehenden Arbeit', Physica XII (1946), 509.

[14] MacCallum, C.J., Invariant Imbedding and Wave Propagation in Inhomogeneous Media, Sandia Corporation, Sc-4669 (RR) (Nov. 1962).

[15] Bellman, R., R. Vasudevan, and S. Ueno, 'On the Matrix Riccati Equation of Transport Processes', Journal of Mathematical Analysis and Application 44 (1973), 472.

[16] Lippman, B.A. and J. Schwinger, Physical Review 79 (1950), 469.

[17] Allen Jr., R.C. and G.M. Wing, 'A Method of Computing Eigenvalues of Certain Schrödinger-like Equations', Journal of Mathematical Analysis and Application 15 (1966), 340.

CHAPTER IX

OPERATOR TECHNIQUES

In the first two sections of this chapter, we will have a brief look at a different approach to the solution of second order linear differential equations based upon operator techniques. Such techniques have been employed widely in physics and other fields of analysis. In quantum mechanics, one is often concerned with the time evolution operator u(t) which satisfies the differential equation of the type

$$\frac{du(t)}{dt} = Au \tag{1}$$

$$u(0) = 1 \tag{2}$$

The Bloch equation, the master or rate equations in chemical kinetics and in statistical mechanics are cases in point. Operator-calculus to obtain the perturbation series for the solution of Equation (1) have been devised in many fields [7, 8, 9]. The solution of Equation (1) can be written as

$$u(t) = T\exp\left(\int_0^t A(\tau)\, d\tau\right). \tag{3}$$

T is called the time ordering operator and if we write

$$u(t) = \sum_{n=0}^{\infty} u_n(t) \tag{4}$$

$$u_0 = 1$$

$$u_n(t) = \int_0^t dt_{n-1} \int_0^{t_{n-1}} dt_{n-2} \cdots \int_0^{t_1} dt_0 A(t_0)A(t_1)\cdots A(t_{n-1}) \tag{5}$$

$$n = 1, 2, \cdots, \infty$$

we have to take care to have the factors in the integrand ordered in time. Also the limits on the integrals are to be ordered. This is important since $A(t_1)$ need not commute with $A(t_2)$ when t_1 and t_2 are different. Because of this, it is not allowed to perform the integration in the exponent. To overcome this Magnus [4] by generalizing the Baker-Campbell-Hausdorff Theorem [2] to the continuous case was able to express u(t) as a series of terms in the exponent. This series can be cut off at any term, preserving the unitarity of the approximate u solution as we shall see later. To derive this we study some well-known operator identities like the BCH series and the Zassenhaus expansions. There are a number of papers relating to these topics and we follow some of the methods described in the paper by Wilcox [1]. These expansions have proved to be of great advantage in many physical problems.

In the later two sections we deal with higher dimensional equations. Imbedding equations for the reflection and transmission functions for the three-dimensional wave equation can be described in analogy with the one-dimensional equations. However, much remains to be studied in this field.

2. THE BAKER-CAMPBELL-HAUSDORFF SERIES

Since the second order equation can be reduced to a first order vector matrix equation of the form

$$\frac{dy}{dt} = Ay \tag{6}$$

various operator techniques are very useful in obtaining the solution y to various orders of accuracy. With this end in view, we derive some of the often used identities by simple procedures and parameter differentiation techniques collected and presented in great detail by Wilcox [1].

To start with we try to express the product of two exponential matrices $e^A e^B$ as e^C. The matrix C can be found to be a series containing A, B, and multiple commutators of A and B. To arrive at this BCH series [3] we use the well-known identity

$$e^A B e^{-A} = B + [A, B] + \frac{1}{2!}[A, [A, B]] + \frac{1}{3!}[A, [A, [A, B]]] + \cdots \tag{7}$$

where [A, B] is the commutator AB - BA. If A and B do not commute this is not zero. A simple proof of Equation (7) can be had if we define

$$f(x) = e^{xA} Be^{-Ax} \tag{8}$$

x being a scalar. Differentiating this with respect to x we have

$$\left.\frac{df}{dx}\right|_{x=0} = \left[A, f(x)\right]\Big|_{x=0} = [A, B], \tag{9}$$

$$\left.\frac{d^2f}{dx^2}\right|_{x=0} = [A, [A, B]] , \tag{10}$$

etc. We know that f(0) = B. Hence, expanding f(x) in a Taylor series around x = 0, we find

$$e^{xA}Be^{-Ax} = f(x) = B + x[A, B] + \frac{x^2}{2!} [A, [A, B]] + \frac{x^3}{3!} [A, [A, [A, B]]] + \cdots . \tag{11}$$

If x = 1 we have the identity (7). Let us now turn to another well-known identity very much used in physics. If the operator H is a function of λ,

$$\frac{\partial}{\partial\lambda} e^{-\beta H(\lambda)} = - \int_0^\beta e^{-(\beta-u)H} \frac{\partial H}{\partial\lambda} e^{-uH} du = F(\beta, \lambda). \tag{12}$$

Both the right-hand side and left-hand side of the equation satisfy the first order linear differential equation

$$\frac{\partial F(\beta, \lambda)}{\partial\beta} + HF(\beta, \lambda) = - \left(\frac{\partial H}{\partial\lambda}\right) e^{-\beta H} \tag{13}$$

with the same initial condition F(0, λ) = 0.

If now in Equation (12) we replace the quantities -H(λ) as z(λ), β by unity and u by (1-s), we have

$$\frac{\partial e^{z(\lambda)}}{\partial\lambda} = \int_0^1 ds\ e^{sz}\,\frac{\partial z}{\partial\lambda}\,e^{-sz}e^{z}. \tag{14}$$

We seek to express z as a power series in λ

$$z = \sum_{n=1}^{\infty} \lambda^n z_n;\ z' = \sum_{n=1}^{\infty} n\lambda^{n-1} z_n \tag{15}$$

such that

$$e^z = e^{\lambda A}\,e^{\lambda B} \tag{16}$$

is identically satisfied (z_n's are to be found in terms of A and B).
Differentiating (16) with respect to λ and utilizing Equation (14), we find

$$\int_0^1 ds\ e^{sz}\,\frac{\partial z}{\partial\lambda}\,e^{-sz} = A + e^{\lambda A}Be^{-\lambda A}. \tag{17}$$

The quantity $e^{\lambda A}Be^{-\lambda A}$ can be expanded using Equation (7)

$$e^{\lambda A}Be^{-\lambda A} = \sum_{j=0}^{\infty} \lambda^j (j!)^{-1}\{A^j, B\}, \tag{18}$$

where the repeated commutator bracket is defined inductively as

$$\begin{aligned} \{A^0, B\} &= B \\ \{A^{n+1}, B\} &= [A, \{A^n, B\}]. \end{aligned} \tag{19}$$

Hence, the integral on the left-hand side of (17) can be written in view of Equation (15) as

$$\begin{aligned} \int_0^1 ds\ e^{sz}\,\frac{\partial z}{\partial\lambda}\,e^{-sz} &= \sum_{k=0}^{\infty} \frac{\{z^k, z'\}}{(k+1)!} = \\ &= \sum_{k=0}^{\infty} \left\{\frac{1}{(k+1)!}\left(\sum_{m=1}^{\infty} \lambda^m z_m\right)^k, \left(\sum_{n=1}^{\infty} n\lambda^{n-1} z_n\right)\right\}. \end{aligned} \tag{20}$$

$$= A + \sum_{j=0}^{\infty} \lambda^j (j!)^{-1}\{A^j, B\}. \tag{21}$$

Equating coefficients of λ^j in expressions (20) and (21) since they are identical, yields various terms of the expansion of z. For j = 0, we have

$$z_1 = A + B. \tag{22}$$

For j = 1, we get, using notation of Equation (18)

$$\{z^0, 2z_2\} + \frac{1}{2}\{z_1, z_1\} = \{A, B\} \tag{23}$$

and

$$z_2 = \frac{1}{2}[A, B]. \tag{24}$$

For j = 2, we obtain

$$\{z^0, 3z_3\} + \frac{1}{2}\{z_1, 2z_2\} + \frac{1}{2}\{z_2, z_1\} + \frac{1}{6}\{z_1^2, z_1\} = \frac{1}{2}\{A^2, B\}. \tag{25}$$

Hence, we have, using the values of z_1 and z_2 obtained earlier

$$z_3 = \frac{1}{12}[A, [A, B]] + \frac{1}{12}[[A, B], B]. \tag{26}$$

The successive terms of the z_j series can be found, in principle, though it turns out to be a laborious task. However, one is able to see that when λ = 1, we have the Baker-Campbell-Hausdorff relation

$$e^A e^B = \exp\{A + B + \frac{1}{2}[A, B] + \frac{1}{12}[A, [A, B]] + \frac{1}{12}[[A, B], B] + \cdots\}. \tag{27}$$

The expansion is not unique due to the existence of Jacobi's identity and other such identities

$$[[[A, B], C], D] + [[[B, C], D], A] + [[[C, D], A], B] + [[[D, A], B], C] = [[A, C], [B, D]] \tag{28}$$

$$[[A, B], C] + [[C, A], B] + [[B, C], A] = 0. \tag{29}$$

It may happen, however, that the multiple commutator in expression (27) can vanish at a certain stage. Then, it proves to be a useful tool in many applications.

As dual to the BCH series, we have the Zassenhaus formula which expresses $e^{\lambda(A+B)}$ as an infinite product of exponential operators [17]

$$e^{\lambda(A+B)} = e^{\lambda A} e^{\lambda B} e^{\lambda^2 C_2} e^{\lambda^3 C_3}. \tag{30}$$

Differentiating both sides of (30) with respect to λ and multiplying both sides from the right by

$$e^{-\lambda(A+B)} = \cdots e^{-\lambda^3 C_3} e^{-\lambda^2 C_2} e^{-\lambda B} e^{-\lambda A} \tag{31}$$

we have

$$A + B = A + e^{\lambda A} B e^{-\lambda A} + e^{\lambda A} e^{\lambda B} (2\lambda C_2) e^{-\lambda B} e^{-\lambda A} + \cdots . \tag{32}$$

The terms $e^{\lambda A} B e^{-\lambda A}$, etc., on the right-hand side of (32) can be expanded by means of Equation (7). Hence, we have

$$B = \sum_{n=0}^{\infty} \frac{\lambda^n}{n!} \left\{A^n, B\right\} + 2\lambda \sum_{m=0}^{\infty} \sum_{n=0}^{\infty} \frac{\lambda^{n+m}}{m!n!} \left\{A^m, B^n, C_2\right\} + 3\lambda^2 \sum_{k=0}^{\infty} \sum_{m=0}^{\infty} \sum_{n=0}^{\infty} \frac{\lambda^{k+m+n}}{k!m!n!} \left\{A^k, B^m, C_2^n, C_3\right\} + \cdots . \tag{33}$$

The quantities $\left\{A^n, B\right\}$ have been defined in Equation (19) while $\left\{A^m, B^n, C_2\right\}$ are defined inductively as

$$\left\{A^0, B^n, C_2\right\} = \left\{B^n, C_2\right\} \tag{34}$$

$$\left\{A^{m+1}, B^n, C_2\right\} = \left[A, \left\{A^m, B^n, C_2\right\}\right]. \tag{35}$$

High order repeated commutators are similarly defined. The coefficients of λ^j for $j \neq 0$ must vanish on the right-hand side of Equation (32). Hence, for $j = 1$, we obtain

$$C_2 = -\frac{1}{2}[A, B]. \tag{36}$$

Setting $j = 2$, we have

$$\begin{aligned} 0 = \frac{1}{2}\left\{A^2, B\right\} + 2\left\{A, B^0, C_2\right\} + 2\left\{A^0, B, C_2\right\} + \\ + c\left\{A^0, B^0, C_2^0, C_3\right\}. \end{aligned} \tag{37}$$

Hence, using (36) we have

$$C_3 = \frac{1}{3}[B, [A, B]] + \frac{1}{6}[A, [A, B]]. \tag{38}$$

These results could also be obtained by a simpler method in the following fashion. Let

$$X = e^{(A+B)\lambda} \tag{39}$$

$$\frac{dX}{d\lambda} = (A+B)X, \quad X(0) = 1. \tag{40}$$

Set

$$\begin{aligned} X &= e^{A\lambda}Y \\ Y &= e^{B\lambda}W \end{aligned} \tag{41}$$

Then we find the differential equation for w as

$$\frac{dW}{d\lambda} = (e^{-B\lambda}e^{-A\lambda}Be^{A\lambda}e^{B\lambda} - B)W. \tag{42}$$

Utilizing the expansion for $e^{-A\lambda}Be^{A\lambda}$ from Equation (7) we find that

$$\frac{dW}{d\lambda} = \lambda(BA - AB)W + O(\lambda^2). \tag{43}$$

Hence, comparing Equation (30), with $X = e^{(A+B)\lambda} = e^{A\lambda}e^{B\lambda}W$ of Equation (41) we have the result for this order

$$C_2 = \frac{1}{2}[B, A] \tag{44}$$

This is seen if we integrate Equation (43) and obtain $W = e^{(\lambda^2/2)[B, A]}$ up to this order. The successive quantities C_3, C_4, etc., can be found by this method and also a general recurrence relation.

3. THE MAGNUS EXPANSION

In many branches of mathematical physics we deal with an evolution equation of the type

$$\frac{du(t, t_0)}{dt} = A(t)u(t, t_0), \tag{45}$$

with $u(t_0, t_0) = 1$. Let us assume t_0 to be equal to zero and instead of the Dyson expansion where time ordering is involved, we go to the Magnus expansion for $u(t)$, which is the continuous analogue of the BCH expansion. We give the proof presented in Reference [1]. We introduce an iteration parameter in Equation (45).

$$\frac{d}{dt} u(\lambda, t) = \lambda A(t)u(\lambda, t) \tag{46}$$

with $u(\lambda, 0) = 1$ and call $u(\lambda, t) = y(\lambda, t)$ such that

$$u(1, t) = y(t) = u(t). \tag{47}$$

Let us assume a solution of (45) of the form

$$y(\lambda, t) = \exp \Omega(\lambda, t), \tag{48}$$

where

$$\Omega(\lambda, t) = \sum_{n=1}^{\infty} \lambda^n \Delta_n(t). \tag{49}$$

If we now recall Equation (14) and do the differentiation of $z(\lambda, t)$ with respect to t instead of with respect to λ, and replace $z(\lambda, t)$ by $\Omega(\lambda, t)$, we obtain in view of Equations (48) and (45)

$$\int_0^1 ds\; e^{s\,(\lambda,t)} \frac{d\Omega}{dt}\, e^{-s\Omega(\lambda,t)} = \lambda A(t). \tag{50}$$

Substituting Equation (49) for $\Omega(\lambda, t)$ and noting that $\int_0^1 s^j\, ds = 1/j+1$, we arrive at

$$\left\{\sum_{k=0}^{\infty} \frac{1}{(k+1)!}\left(\sum_{n=1}^{\infty} \lambda^n \Delta_n(t)\right)^k, \left(\sum_{m=1}^{\infty} \lambda^m \dot{\Delta}_m(t)\right)\right\} = \lambda A(t),$$
$$\dot{\Delta}_m(t) = \frac{d\Delta_m(t)}{dt}. \tag{51}$$

The curly bracket stands for the repeated commutator operations indicated earlier. Equating the coefficients of λ^j on both sides of Equation (51) we obtain recursively equations for Δ_1, Δ_2, ..., etc. Setting $j = 1$

$$\dot{\Delta}_1 = A(t); \qquad \Delta_1 = \int_0^t A(\tau)\, d\tau. \tag{52}$$

For $j = 2$, we have

$$\dot{\Delta}_2 + \tfrac{1}{2}[\Delta_1(t), \dot{\Delta}_1(t)] = 0. \tag{53}$$

Hence

$$\Delta_2(t) = \frac{1}{2}\int_0^t d\sigma \int_0^\sigma d\tau [A(\sigma), A(\tau)]. \tag{54}$$

Similarly, we have

$$\Delta_3(t) = \frac{1}{6}\int_0^t dt_1 \int_0^{t_1} dt_2 \int_0^{t_2} dt_3 \left\{\Big[[A_1, A_2], A_3\Big] + \right.$$
$$\left. + \Big[[A_3, A_2], A_1\Big]\right\}, \tag{55}$$

where $A_1 = A(t_1)$, etc. Thus higher order terms can be found and the solution y(t) can now be obtained by setting $\lambda = 1$.

Thus

$$u(t) = T \exp \int_0^t A(\tau)\, d\tau$$

$$= \exp \left\{ \int_0^t A(\tau)\, d\tau + \frac{1}{2} \int_0^t dt_1 \int_0^{t_1} dt_2 [A(t_1), A(t_2)] + \right. \tag{56}$$

$$+ \frac{1}{6} \int_0^t dt_1 \int_0^{t_1} dt_2 \int_0^{t_2} dt_3 \left[\left[[(A(t_1), A(t_2)], A(t_3)\right] + \right.$$

$$\left. + \left[[A(t_3), A(t_2)], A(t_1)\right]\right] + \ldots \text{ etc.} \Bigg\}.$$

To start with A(t) in the evolution operator equation is Hermitian and each term in the exponential remains Hermitian. Hence, truncating the expansion at any place yields a unitary approximation for u(t). This expansion has been used in many physical problems such as, for example, the time development operator arising in some cases in nuclear-magnetic resonance spin dynamics with a time dependent Hamiltonian [5].

Instead of expressing $y(\lambda, t)$ as $\exp \Omega(\lambda, t)$ we can write it as

$$u(\lambda, t) = e^{\lambda S_1} e^{\lambda^2 S_2} e^{\lambda^3 S_3} \ldots \text{ etc.} \tag{57}$$

and we can determine the quantities S_n by substituting (57) into Equation (46). We multiply from the right by $[u(\lambda, t)]^{-1}$ and expanding in terms of repeated commutators, we have

$$\sum_{n=0}^{\infty} \frac{\lambda^{n+1}}{(n+1)!} \left\{ S_1^n, \dot{S}_1 \right\} + \sum_{m=0}^{\infty} \sum_{n=0}^{\infty} \frac{\lambda^{m+2n+2}}{m!(n+1)!} \left\{ S_1^m, S_2^n, \dot{S}_2 \right\} + \tag{58}$$

$$+ \cdots = \lambda A(t).$$

Equating the coefficients of λ^j on both sides, we have

$$S_1(t) = \int_0^t A(\tau)\, d\tau \tag{59}$$

$$S_2(t) = \frac{1}{2}\int_0^t d\sigma \int_0^\sigma d\tau [A(\sigma), A(\tau)] \tag{60}$$

$$S_3(t) = \frac{1}{3}\int_0^t dt_1 \int_0^{t_1} dt_2 \int_0^{t_2} dt_3 \Big\{[A_2, [A_3, A_1]] + + [A_3, [A_2, A_1]]\Big\} \cdots . \tag{61}$$

Hence, with $\lambda = 1$, $y(t)$ can now be expressed as an infinite product of the exponentials. This is called the Fer formula [6] and is the continuous analogue of the Zassenhas expansion given earlier.

If in addition A can be expressed as a linear combination of matrices that represent the elements of a Lie algebra, i.e.,

$$A = a_1(t)x_1 + a_2(t)x_2 + \cdots + a_n(t)x_n, \tag{62}$$

where $x_1, x_2, \cdots, x_n$ are the elements or representations of Lie algebra. Then the solution of the form $y(t) = \exp \Omega(t)$ can be obtained in a closed form as a product

$$y(t) = \exp [g_1(t)x_1] \exp [g_2(t)x_2] \cdots [\exp g_n(t)x_n] \tag{63}$$

This is implied by the repeated commutator form for Ω and the closure property of the Lie algebra. A very good analysis of these matters is found in the paper by Wilcox [1].

4. HIGHER DIMENSIONAL WAVE EQUATIONS

The three-dimensional wave equation of the form

$$\nabla^2 u + \lambda^2 n^2(\vec{r})u = 0 \quad \lambda >> 1 \tag{64}$$

is a partial differential equation of elliptic type and its solution can still be found for a slowly varying vector $n(\vec{r})$ by introducing the eikonal function $S(r)$ where S is a scalar function of the position $\vec{r}$. From the eikonal solution, we can get as was done in Chapter II the Riccati equation for S (by taking the solution $u = e^{i\lambda S(r)}$),

$$- i \frac{\nabla^2 S}{\lambda} + (\nabla S)^2 - n^2 = 0. \tag{65}$$

If $n(\vec{r})$ is a slowly varying function of its variables then we can approximate by taking $\nabla^2 S = 0$ and Equation (65) reads**

$$(\nabla S)^2 = n^2. \tag{66}$$

If S can be expanded as*

$$S = S_0 + \frac{S_1}{i\lambda} + \frac{S_2}{(i\lambda)^2} + \cdots \tag{67}$$

then we have the set of equations

$$\begin{aligned} &(\nabla S_0)^2 - n^2 = 0 \\ &\nabla^2 S_0 + 2\nabla S_0 \cdot \nabla S_1 = 0 \\ &\nabla^2 S_1 + (\nabla S_1)^2 + 2\nabla S_0 \cdot \nabla S_2 = 0. \end{aligned} \tag{68}$$

We can introduce the vector $\vec{y}_\ell$ in analogy with the one-dimensional case by taking

$$\vec{y}_\ell = \vec{\nabla} S_\ell. \tag{69}$$

---

* This is valid if variation of n is small; i.e., $\left|\frac{\nabla n}{n}\right| << \lambda$.

** Notational indications for scalar, vector and matrix entities can easily be understood from the context. Hence regular notation is not strictly followed in this chapter.

The set of equations for the $\vec{y}_\ell$ will read

$$y_0^2 = n^2 \tag{70}$$

$$\vec{y}_1 \cdot \vec{y}_0 = -\frac{1}{2}(\nabla \cdot y_0) \tag{71}$$

$$\vec{y}_2 \cdot \vec{y}_0 = -\frac{1}{2}[\nabla \cdot \vec{y}_1 + y_1^2] \tag{72}$$

etc. Once we solve this set successively S_n are obtained by simple quadrature. We have not gone into the question of boundary conditions here. Before developing methods by imbedding analysis for three or multi-dimensional cases we will consider a simple two dimensional case.

For two-dimensional equations of the form

$$u_{xx} + u_{yy} \pm a^2(x, y)u = 0 \tag{73}$$

we can effect transformations of the variables x and y in such a way that in terms of the new variables say s and t Equation (73) becomes

$$u_{ss} + u_{tt} \pm u = 0. \tag{74}$$

This is possible only if a(x, y) satisfies certain requirements as shown below. We follow the exposition by Bellman in Reference [10].

In the one-dimensional case we have seen in Chapter II how the Liouville transformation carries over to the equation of the form

$$u'' + a^2(x)u = 0 \tag{75}$$

to the following

$$v'' + [(\pm 1) + b(s)]v = 0. \tag{76}$$

This facilitates arriving at various approximations for the solution if b(s) turns out to be small.

To transform Equation (73) into a more suitable form, we replace the independent variables x and y by two yet unspecified quantities s and t. Equation (73) is converted into the following:

$$u_{ss}(s_x^2 + s_y^2) + u_{tt}(t_x^2 + t_y^2) + 2u_{st}(s_x t_x + s_y t_y) + \\ + u_s(s_{xx} + s_{yy}) + u_t(t_{xx} + t_{yy}) \pm a^2(x, y)u = 0. \tag{77}$$

We wish to determine s and t so that the following relations hold (if we take + sign before a^2 in Equation (77))

$$s_x^2 + s_y^2 = a^2 \tag{78}$$

$$t_x^2 + t_y^2 = a^2 \tag{79}$$

$$s_x t_x + s_y t_y = 0. \tag{80}$$

All these can be satisfied if we have

$$s_x = a \cos \varphi \qquad t_x = a \sin \varphi \tag{81}$$

$$s_y = a \sin \varphi \qquad t_y = -a \cos \varphi. \tag{82}$$

The Jacobian of the transformation is thus $a^2(x, y)$ assuming $a^2 > 0$ for $x > x_0$ and $y > y_0$.

For the transformations (81) and (82) to hold, we must have

$$(s_x)_y = (s_y)_x ; (t_x)_y = (t_y)_x. \tag{83}$$

Equation (83) demands that

$$\varphi_x = \frac{a_y}{a} = \frac{d}{dy} \log a \tag{84}$$

$$\varphi_y = -\frac{a_x}{a} = -\frac{d}{dx} \log a. \tag{85}$$

It follows therefore that φ exists if and only if

$$\frac{d^2}{dx^2}(\log a) + \frac{d^2}{dy^2}(\log a) = 0. \tag{86}$$

In other words, we assume that log a(x, y) is a harmonic function.

If Equation (86) is valid then it is easy to see that

$$s_{xx} + s_{yy} = 0 \tag{87}$$

$$t_{xx} + t_{yy} = 0. \tag{88}$$

Thus, Equation (77) turns out to be

$$u_{ss} + u_{tt} \pm u = 0, \tag{89}$$

where the potential term has become a constant ± 1.

Equation (89) can be solved by separation of variables assuming u = X(s)Y(t) very easily. For more details, see Reference [10].

5. MULTI-DIMENSIONAL IMBEDDING

In this section we show how the imbedding method can be extended to a higher dimensional wave equation or the elliptic partial differential equation with suitable boundary conditions. We describe the method of Maynard and Scott [16] for a simple case.

We consider the equation in three dimensions.

$$\nabla^2\omega(r) + \lambda^2\omega(\vec{r}) = 0 \quad \vec{r} \in R \tag{90}$$

We adopt vector notation and suitable boundary conditions may be obtained by dividing the surface S of the region R into two parts S_1 and S_2 and then requiring

$$\vec{n} \cdot \nabla\omega(r_S) = 0 \quad \text{if} \quad \vec{r}_S \in S_1 \tag{91}$$

$$\vec{n} \cdot \nabla\omega(r_S) = f(\vec{r}_S) \quad \text{if} \quad \vec{r}_S \in S_2\,, \tag{92}$$

where $\vec{n}$ is the unit normal vector to the surface S and $f(r_S)$ is a given function of the vector $\vec{r}_S$. This is in complete analogy with the one-dimensional problems we have treated in earlier chapters. In the one-dimensional case we know that we can form a vector matrix equation of the bivector u and u'.

We will treat $\nabla\omega$ as an independent quantity and define a vector valued function $\vec{S}(\vec{r})$ by the generalized Riccati transformation

$$\nabla\omega(r) = \vec{S}\omega(r) \tag{93}$$

and a scalar valued function $T(\vec{r})$ corresponding to the recovery transformation

$$\omega(\vec{r}'_S) = T(r)\omega(r), \tag{94}$$

where r'_S is an arbitrarily fixed point on S_1. The differential equations for $\vec{S}$ and T functions correspond to the differential equations for the reflection and transmission functions in the one-dimensional case. We are now dealing with vector equations.

Taking the divergence of Equation (93), and using Equation (90) we have

$$[\nabla \cdot \vec{S}(r) + S(r) \cdot S(r) + \lambda^2]\omega(r) = 0 \tag{95}$$

which holds for all $r \in R$. Thus the vector Riccati equation

$$-\nabla \cdot \vec{S}(r) = \lambda^2 + \vec{S}(r) \cdot \vec{S}(r). \tag{96}$$

Also taking the curl of Equation (93), we have

$$[\nabla \times \vec{S}(r) + \vec{S}(r) \times \vec{S}(r)]\omega = 0, \tag{97}$$

since curl of a gradient is zero. This means

$$\nabla \times \vec{S}(r) = 0. \tag{98}$$

From the boundary condition (91) and (92) we know that

$$n \cdot \vec{S}(r_S) = 0 \quad r_S \in S_1. \tag{99}$$

Thus the imbedding equations for $\vec{S}(r)$ are complete with the boundary conditions on one part S_1 of the surface S. We want the equations for the function T(r) and also to obtain the internal values. To this end we apply the gradient operator to Equation (94) and obtain

$$0 = \omega(r)\nabla T(r) + T(r)[\vec{S}(r)\omega(r)], \tag{100}$$

where the left side is zero since $r_S' \in S_1$. Hence, we have

$$\nabla T(r) + \vec{S}(r)T(r) = 0. \tag{101}$$

The initial condition for this equation is

$$T(r_S') = 1, \tag{102}$$

since Equation (94) must hold for any arbitrary point in the region $r_S' \in S_1$. We can integrate Equations (101) and (96) at this stage. But we have not made use of the other boundary condition (92). We find for the points on S_2 using (92),

$$n \cdot \nabla\omega(r_S'') = n \cdot \vec{S}(r_S'')\omega(r_S''); \quad r_S'' \in S_2 \tag{103}$$

$$\omega(r_S'') = \frac{f(r_S'')}{n \cdot \vec{S}(r_S'')}; \; r_S'' \in S_2. \tag{104}$$

Inserting this into Equation (94) for r_S'' in S_1 yields

$$\omega(r_S') = T(r_S')\omega(r_S'') = \frac{T(r_S'')f(r_S'')}{n \cdot \vec{S}(r_S'')}\; ; \quad \begin{array}{l} r_S' \in S_1 \\ r_S'' \in S_2 \end{array} \tag{105}$$

Now that we know $\omega(r_S')$ we can use Equation (94) to solve for $\omega(r)$ for any $r \in R$

$$\omega(r) = \frac{T(r_S'')f(r_S'')}{T(r)n \cdot \vec{S}(r_S'')}\,. \tag{106}$$

For $\omega(r)$ to be well defined for all r the denominator of (106) should not become zero. This can be guaranteed by restricting the region R suitably. We have sketched briefly the main trenads of the arguments of the reference [16]. For further developments, the reader should look into the literature cited.

6. HIGHER ORDER EQUATIONS

The results of the previous chapters can be extended to include linear differential equations of the nth order [11] such as

$$u^N + P_1(x)u^{N-1} + \cdots P_N u(x) = 0, \tag{107}$$

where we suppose that

$$P_1(x) = p_i + q_i(x) \quad i = 1, 2, \cdots, N \tag{108}$$

where p_i are constant and $q_i(x)$ are small and $q_i(x) \to 0$ as $x \to \infty$. Let us suppose that the roots of the polynomial equation

$$r^N + p_1 r^{N-1} + \cdots p_N = 0 \tag{109}$$

which we shall call $\lambda_1, \lambda_2, \ldots, \lambda_N$ are distinct. Then the roots of the equation

$$r^N + P_1(x)r^{N-1} + P_2(x)r^{N-2} + \cdots P_N(x) = 0 \tag{110}$$

which we shall call $\lambda_1(x), \lambda_2(x) \ldots \lambda_N(x)$ are distinct for $x \geq 0$ if the $|q_i(x)|$ are sufficiently small - which we assume.

Let $q_i(x)$ be identically zero for $x \leq 0$. By a wave we shall now mean a solution of (107) for $x < 0$ having the form

$$u = e^{\lambda_i x}. \tag{111}$$

If the medium for $x > 0$ is homogeneous which is to say that if $P_i(x)$ are constants; for $x > 0$, we can ask for analogues of the reflected and transmitted waves as in previous chapters.

Let λ_i' be the natural continuation of λ_i as x passes through zero. Then we look for a solution of (107) of the form

$$u = c_i e^{\lambda_i' x} \tag{112}$$

in $x > 0$ and of the form

$$u = e^{\lambda_i x} + \sum_{j \neq i} c_j e^{\lambda_j x} \tag{113}$$

for $x < 0$.

The term $e^{\lambda_i' x}$ we can call the transmitted wave and the sum $\Sigma_{j \neq i}\, c_j e^{\lambda_j x}$ a reflected wave. As was done earlier, the N coefficients $c_1, c_2, \ldots, c_N$ are to be determined by the continuity conditions, i.e., continuity of u and its first (N-1) derivatives as x approaches zero from both sides. These relations are

$$
\begin{aligned}
c_i &= 1 + \sum_{j \neq i} c_j \\
\lambda_i' c_i &= \lambda_i + \sum_{j \neq i} \lambda_j c_j \\
&\cdots \\
(\lambda_i')^{N-1} c_i &= \lambda_i^{N-1} + \sum_{j \neq i} \lambda_j^{N-1} c_j .
\end{aligned}
\tag{114}
$$

The above N equations corresponding to each i can be solved since the determinant of each of the N × N matrix equation is a Vandermonde's determinant and it does not vanish, due to the assumption that the λ_i's are distinct. They yield a unique solution for the c_i since for each i there exists c_j's which correspond to that particular i. We change our notation slightly and let $c_j = c_{ij}$ $j \neq i$ but leave c_i unchanged. We can now call the oncoming wave vector

$$
y = \begin{bmatrix} e^{\lambda_1 x} \\ e^{\lambda_2 x} \\ \vdots \\ e^{\lambda_N x} \end{bmatrix}
\tag{115}
$$

and the transmitted wave vector

$$\vec{y}_1 = B \begin{bmatrix} e^{\lambda'_1 x} \\ e^{\lambda'_2 x} \\ \vdots \\ e^{\lambda'_X x} \end{bmatrix} , \tag{116}$$

where the matrix $\underline{B}$ is a diagonal matrix

$$B = \begin{bmatrix} c_1 & 0 & 0 & 0 & \cdots & 0 \\ 0 & c_2 & . & . & \cdots & . \\ 0 & . & . & . & \cdots & . \\ 0 & & & & & c_N \end{bmatrix} \tag{117}$$

and the reflected wave for $x < 0$

$$\vec{y}_2 = \omega \begin{bmatrix} e^{\lambda_1 x} \\ e^{\lambda_2 x} \\ \vdots \\ e^{\lambda_N x} \end{bmatrix} , \tag{118}$$

ω matrix is represented as

$$\omega = \begin{bmatrix} 0 & c_{12} & \cdots & c_{1N} \\ c_{21} & 0 & & c_{2N} \\ c_{N1} & & & 0 \end{bmatrix} . \tag{119}$$

These B and ω matrices which are the transmission and reflection matrices can be found as functions of the point of incidence of the wave. Using the principle of localization and other invariance concepts, we can arrive at the differential equations for the reflection function matrix and transmission function matrix for a rod extending from $x = 0$ to $x = z$. We can then construct the internal solutions and obtain the WKB and higher approximations as a Bremmer series. Also, under various assumptions we can find the asymptotic solutions. As an illustrative example we will consider a third order equation of the form

$$y^{(3)} + 3p_1(x)y^{(2)} + 3p_2(x)y^{(1)} + p_3(x)y = 0. \tag{120}$$

The superscript denotes the number of times y has been differentiated. We can under suitable conditions transform this equation by changing the variable x to $\xi(x)$ to the form

$$y''' + 3\bar{p}_1 y'' + 3\bar{p}_2 y' + \bar{p}_3 y = 0. \tag{121}$$

The primes indicate differentiation with respect to ξ and if another transformation

$$y = \bar{y} \exp - \int \bar{p}_1 \, d\xi \tag{122}$$

is effected, Equation (121) becomes

$$\bar{y}''' + 3P_2\bar{y} + P_3\bar{y} = 0. \tag{123}$$

If the ξ transformation is so chosen that $P_2 = 0$ we can always arrive at the equation

$$\bar{y}''' + P_3\bar{y} = 0. \tag{124}$$

The reader is referred to literature in References [12], [13] for details of this transformation.

Let us now work with the third order equation of the form

$$\frac{d^3\Psi}{dx^3} + \lambda^3 q^3(x)\Psi = 0. \tag{125}$$

This can be written in the matrix form

$$\frac{d}{dx}\vec{\Psi} = Q\Psi \tag{126}$$

with

$$\vec{\Psi} = \begin{bmatrix} \Psi \\ \Psi' \\ \Psi'' \end{bmatrix} \quad ; \quad Q = \begin{bmatrix} 0 & 1 & 0 \\ 0 & 0 & 1 \\ -\lambda^3 q^3 & 0 & 0 \end{bmatrix} \tag{127}$$

The diagonalizing matrix for Q is T

$$T = \begin{pmatrix} 1 & 1 & 1 \\ -\lambda q & -\lambda\omega q & -\lambda\omega^2 q \\ \lambda^2 q^2 & \lambda^2\omega^2 q^2 & \lambda^2\omega q^2 \end{pmatrix} \tag{128}$$

(ω being the cube root of unity $\omega = \exp(2\pi i/3)$, and the inverse of the T matrix is

$$T^{-1} = \frac{1}{3}\begin{pmatrix} 1 & -\frac{1}{\lambda q} & \frac{1}{\lambda^2 q^2} \\ 1 & -\frac{\omega^2}{\lambda q} & \frac{\omega}{\lambda^2 q^2} \\ 1 & -\frac{\omega}{\lambda q} & \frac{\omega^2}{\lambda^2 q^2} \end{pmatrix} , \tag{129}$$

Hence, Equation (126) after the transformation

$$\vec{\Psi} = T\vec{y}_1 \tag{130}$$

becomes

$$\frac{d\vec{y}_1}{dx} = -\lambda q C\,\vec{y}_1 - \frac{q'}{q} I\vec{y}_1 - - \frac{q'}{3q}\{(\omega^2-1)A + (\omega-1)A^2\}, \tag{131}$$

where the C and A matrices are representations of the elements of the generalized Clifford algebra defined by

$$AC = \omega CA \tag{132}$$

and obeying the conditions

$$A^3 = C^3 = 1 \tag{133}$$

$$A = \begin{bmatrix} 0 & 1 & 0 \\ 0 & 0 & 1 \\ 1 & 0 & 0 \end{bmatrix} \quad ; \quad C = \begin{bmatrix} 1 & 0 & 0 \\ 0 & \omega & 0 \\ 0 & 0 & \omega^2 \end{bmatrix} \tag{134}$$

ω is the cube root of unity. Details of the properties of these elements and their applications in physics can be found in the book by Ramakrishnan [14].

If we now make another transformation

$$\vec{y}_1 = q^{-1}\vec{y}_2 \tag{135}$$

we obtain

$$\frac{d\vec{y}_2}{dx} = -\lambda q C y_2 - \frac{q'}{3q}\{(\omega^2 - 1)A + (\omega - 1)A^2\}. \tag{136}$$

If λ is large, the second term in the right-hand side of Equation (136) is smaller by a factor $1/\lambda$. Hence, as an approximation it can be ignored. If we are left with only the first term on the right-hand side of (136), since C is a diagonal matrix, the vector $\vec{y}_2$ is immediately obtained and represents the WKB like approximation to the solution of the third order equation. If u_1, u_2, and u_3 are the components of $\vec{y}_2$ a set of three coupled integral equations can be obtained by directly integrating (136). This will be analogous to the two coupled equations for the second order equation of Chapter IV. Hence, the Bremmer series type solutions are obvious for given boundary conditions. For further sophistications of this analysis the reader is referred to the preprint [15].

REFERENCES

[1] Wilcox, R.M., 'Exponential Operation and Parameter Differentiation in Quantum Physics', Journal of Mathematical Physics 8 (1966), 962.

[2] Baker, H.F., 'On the integration of Linear Differential

Equations', Proceedings of the London Mathematical Society, 34 (1902) 347; 35 (1903), 333; Series 2 (1904), 293; L. Campbell, Proceedings of the London Mathematical Society 29 (1898), 14.

3 Weiss, H. and A.A. Maradudin, 'Baker-Housedarf Formula and a Problem in Crystal Physics', Journal of Mathematical Physics 3 (1962), 771.

4 Magnus, W., 'On the Exponential Solution of Differential Equations for a Linear Operator', Communications of Pure and Applied Mathematics 7 (1954), 649.

5 Evans, W.A.B., 'On Some Applications of Magnus: Expansion in Nuclear Magnetic Resonance', Annals of Physics 48 (1968), 72.

6 Fer, F., Bulletin Classe - Science Academy Royal Belgium 44 (1958), 818.

7 Feynman, R.P., 'An Operator Calculus Having Applications in Quantum Electrodynamics', Physics Review 84 (1951), 108.

8 Per-Olov, Lowdin, 'Studies in Perturbation Theory IV: Solutions of Eigenvalue Problem by Projection Operator Formalism', Journal of Mathematical Physics 3 (1962), 969.

9 Chen, K., 'Decomposition of Differential Equations', Math. Annalen, 146 (1962), 263.

10 Bellman, R., 'On a Liouville Transformation of $u_{xx} + u_{yy} + a^2(x, y)u = 0$', Boll. Dlunione Matematico 13 (1958), 535.

11 Bellman, R. and R. Kalaba, 'Functional Equations, Wave Propagation and Invariant Imbedding', Journal of Math. and Mech. 8 (1959), 688.

12 Wilczynki, E.J., Projective Differential Geometry of Curves and Surfaces, Chelsea Publishing Co., 1905.

13 Fosyth, A.R., Theory of Differential Equations, Vols. III and IV, Dover Publications, New York, 1959.

14 Ramakrishnan, A., L-Matrix Theory or the Grammar of Dirac Matrices, Tata-McGraw-Hill Book Co., Bombay, India, 1972.

15 Vasudevan, R. and A.K. Ganguli, MATSCIENCE Preprint, 1975.

16 Maynard, C.W. and M.R. Scott, 'Invariant Imbedding of Linear Partial Differential Equations for a Generalized Riccati Transformation', Journal of Mathematical Analysis and Applications 36 (1971) 432.

17 Sudarshan, E.C.G., in Brandeis Summer Institute Lectures in Theoretical Physics, W.A. Benjamin, New York, 1962, 181.

CHAPTER X

VARIATIONAL PRINCIPLES

1. INTRODUCTION

In this chapter we deal with variational principles of great universality and wide scope in solving the equations of the type we were concerned with in previous chapters. The minimum hypothesis to explain natural phenomena seems to have been present in the concepts advocated by Aristotle, Hero (c. 125 BC) and other early thinkers whose philosophy was based on the principle that 'Nature does nothing in vain'. Such 'economy' postulates in the hands of later scientists acted as foundation for the development of minimum principles such as the Maupertian and Fermat's principles. If one scrutinizes the chronological stages in the evolution of minimum principles one can obtain a profound insight into the continuous transformation of a metaphysical canon into an exact natural law.

Starting with the work of Euler and Lagrange physicists have succeeded in formulating laws of elasticity, hydrodynamics, electrodynamics, and even Einstein's law of gravitation as variational principles. The recent formulations of laws of quantum mechanics and quantum field theory can be expressed in terms of action principles.

In the second section we describe the Bubnow-Galerkin technique which is based on a simple idea, to obtain the solution of differential equations by minimizing a quadratic functional. Then we go on to describe a method known as the Rayleigh-Ritz method which is also applied to characteristic value problems. We also briefly mention a method of obtaining upper and lower bounds for the functional, based on the concept of duality. In the last section we describe how the variational methods can be handled by dynamic programming techniques developed by Bellman and others.

2. BUBNOV-GALERKIN METHOD

One of the major approximation techniques of modern applied mathematics of great analytical and computational significance is the Bubnov-Galerkin technique which can be applied to obtaining solutions of the differential equation of the form

$$\frac{du}{dt} = g(u); \quad 0 \leq t \leq T \tag{1}$$

subject to initial or boundary values [1, 2]. This method applied to linear equations leads to the 'method of moments'. This method has also been applied to nonlinear equations with great success [3].

Let Equation (1) have an initial condition, say

$$u(0) = c. \tag{2}$$

In case g(u) is a nonlinear function of u, a vector of dimension N, the analytical treatment is intractable. Let us form a functional J(u) defined as

$$J(u) = \int_0^T (u' - g(u), u' - g(u))\, dt. \tag{3}$$

Let us minimize J(u) over all u for which the functional exists subject to the condition the u(0) = c. If u is subject to two point or multipoint boundary conditions, these conditions can be imposed on the trial functions. To obtain the approximate solution to this variational problem, we restrict the candidates to those functions of the type

$$x = y(a, t); \qquad y(a, 0) = c \tag{4}$$

where y(a, t) is an element of the conveniently chosen family of functions dependent on a k dimensional vector a. Then we can write

$$J[y(a, t)] = h(a) = h(a_1, a_2, \cdots, a_k). \tag{5}$$

Thus minimization involves the minimization of a function of k variables, a problem within the domain of classical calculus approachable by nonlinear programming algorithms or by various search techniques of a sophisticated nature. Even this lack of determining the correct values of $a_1, a_2, \cdots, a_k$ is difficult if k is large. A reasonable initial approximation is necessary to start with.

It has been proven in Reference [1] that if g(u) is analytic in u and if y is a function such that

$$\int_0^T [y' - g(y),\ y' - g(y)]dt \leq \varepsilon^2$$

then, provided ε is sufficiently small, we have

$$\|u - y\|_M \leq k_1\varepsilon,$$

where k_1 is a constant independent of ε which depends on the solution of (1). Here we use the norm

$$\|u - y\|_M = \max_{0 \leq t \leq T} \|u - y\| .$$

For a problem with boundary conditions of the type

$$x'' = g(x); \quad x(0) = c \quad x'(T) = 0 \tag{6}$$

we minimize

$$J(x) = \int_0^T [x'' - g(x),\ x'' - g(x)]\ dt \tag{7}$$

using a trial function of finite number of parameters,

$$x \approx Y = \Sigma a_k y^k , \tag{8}$$

where y^k are a specified known set of functions with $Y(0) = c$ and $Y'(T) = 0$. If for some reason it is inconvenient to impose the condition $x'(T) = 0$, we replace $J(x)$ by $J(x) + \lambda\|x'(T)\|$ where $\lambda \geq 1$.

Nonlinear equations of the type

$$u'' + u + \varepsilon u^3 = 0; \quad u(0) = 1 \quad u'(0) = 0 \tag{9}$$

which describe the oscillations of a nonlinear spring can be tackled by this method. If $\varepsilon = 0$, the solutions are the well-known periodic solutions. Consider the functional

$$J(u) = \int_0^\pi (u'' + u + \varepsilon u^3)^2\ dt. \tag{10}$$

The effect of the nonlinearity is to change both the period and the amplitude of the oscillations. Ignoring the change in the amplitude, which turns out to be a smaller effect, we can use a trial function

$$u \approx \cos \omega t, \tag{11}$$

where we can take

$$\omega = 1 + b_1\varepsilon + b_2\varepsilon^2. \tag{12}$$

If we stop with the first order term in ε, we obtain, on minimizing J with respect to b_1, a value for b_1 given by

$$b_1 = \frac{3}{8} \tag{13}$$

which should be compared with the solutions obtained by straightforward perturbation solutions aided by 'tucking in' techniques [1].

In the case of the linear functional equation the Bubnov-Galerkin technique is very effective, since after minimization of the parameter we obtain a system of linear algebraic equations, called the 'moment' equations. For example, if we consider the equation

$$L(u) = g, \tag{14}$$

where L is a linear scalar operator, we minimize the functional

$$J(u) = \int_0^T (L(u) - g)^2\, dt \tag{15}$$

with trial functions of the type

$$u = \Sigma a_k\, u_k, \tag{16}$$

u_k's are prescribed functions obeying the appropriate end conditions. Minimizing over a_k we obtain a system of linear equations

$$m_i = \int \left(\sum_{k=1}^{N} a_k L(u_k) - g\right) L(u_i)\, dt = 0, \tag{17}$$

m_i are the moments of the function $\left(\Sigma_k a_k L(u_k) - g\right)$ with respect to the sequence of functions $\{L(u_i)\}$. The solutions of this system of equations for a_k's depend on the facility with which we can calculate the integrals

$$b_{ij} = \int_0^T L(u_i)L(u_j)\, dt \tag{18}$$

and the conditioning of the matrix $B = b_{ij}$. The conditioning of B often dictates the choice of the function $\{u_k\}$.

If we analyze the above method closely we need not have restricted ourselves to the particular type of moment euqations arrived at in (17). We could have used another convenient choice of sequence of functions $\{v_n\}$ and determine the coefficients by the requirement

$$\sum_{k=1}^{N} a_k \int_0^T L(u_k)v_i\, dt - \int_0^T g\, v_i\, dt = 0 \qquad i = 1, 2, \ldots, N. \tag{19}$$

Suppose $\{L(u_k)\}$ and $\{v_k\}$ are biorthogonal, then we have

$$a_i = \int_0^T g\, v_i\, dt \qquad i = 1, 2, \ldots, N. \tag{20}$$

Hence, we have

$$u^{(N)} = \sum_{k=1}^{N} a_k u_k = \sum_{k=1}^{N} \left(\int g\, v_k\, dt\right) u_k. \tag{21}$$

This is the N-th partial sum of the series for the expansion of the solution of $L(u) = g$, upon setting

$$u = \sum_{n=1}^{\infty} a_n u_m.$$

Hence, under reasonable assumptions concerning the sequences $\{u_n\}$ and $\{v_n\}$ we can expect that the sequence of approximations obtained in this fashion converge at least in the mean square sense.

The Bubnov-Galerkin method may be made rigorous by using some results of stability theory; see Reference [15] .

3. THE RAYLEIGH-RITZ METHOD

As a variational technique, this is a powerful and elegant method to obtain approximate analytic and numerical solutions of functional equations. Let us take a quadratic functional of the form

$$J(u) = \int_0^T g(u, u')\, dt \tag{22}$$

and minimize this over a class of functions {u} subject to an initial condition say $u(0) = c$, or with initial and terminal conditions. We consider the variation of u constructed in the following way. Let $\eta(t)$ be an admissible function defined in the closed interval [0, T]. For sufficiently small absolute values of the parameter ε, the functions

$$\bar{u} = u + \varepsilon\eta \tag{23}$$

lie in the neighborhood of u. We say $\delta u = \varepsilon\eta$ is a variation of the function u. We compute the J at $\bar{u} = u + \varepsilon\eta$

$$J(u + \varepsilon\eta) = \int_0^T g(u + \varepsilon\eta, u' + \varepsilon\eta')\, dt. \tag{25}$$

If $\bar{u}$ has the same condition at $t = 0$ as u, then $\eta(0) = 0$. A necessary condition for the minimum is given by

$$\frac{d}{d\varepsilon} J(u + \varepsilon\eta)\bigg|_{\varepsilon=0} = 0. \tag{26}$$

Differentiating under the integral sign we obtain

$$\int_0^T (g_u\eta + g_{u'}\eta')\, dt = 0. \tag{27}$$

Integrating the second term in (27) by parts, we obtain

$$\int_0^T \left(g_u - \frac{d}{dt} g_{u'}\right)\eta\, dt = 0 \tag{28}$$

valid for all trial functions η defined earlier. Hence, we arrive at the Euler equation

$$\frac{d}{dt} g_{u'} - g_u = 0 \tag{29}$$

subject to the two point boundary conditions that u has been assigned. If u has free boundary conditions at T, η need not be equated to 0 at T, hence, we must have the condition*

$$g_{u'}(T) = 0 \tag{30}$$

added as a terminal condition at T for the Euler equation.

In expanded form, Equation (29) will read as

$$g_{u'u'}u'' + g_{u'u}u' + g_{u't} - g_u = 0 \tag{31}$$

if g is a function of u, u' and t. In order to solve (31) for the higher derivative u" we must also have the condition

$$g_{u'u'} \neq 0 \tag{32}$$

which is known as the Legendre condition.

With the foregoing in mind, it is reasonable to think of reversing the process, starting with Equation (31) as the original object of our attention. Can we use its connection with the minimization of J(u) to obtain an analytic and computational hold on its solution? Once the problem has been transformed into that of minimizing the functional J(u) we can proceed as we did in the case of the Bubnov-Galerkin method to use a number of approximations.

The Euler Equation (31) together with the boundary conditions should be first proven to have a solution which exists and is unique. This can be done and once we have this solution we should also demonstrate that this solution furnishes the absolute minimum of J(u) over a suitable class of functions. Armed with these procedures, we can fruitfully use various computational estimates of the accuracy of the approximations.

Let us take the one dimensional case where the functional has the form

---

* This is not necessary if u(T) = 0, called the free boundary condition.

$$J(u) = \int_0^T (u'^2 + \varphi(t)u^2)\,dt \tag{33}$$

with u(0) = c. Let v be a function such that $v' \varepsilon L^2(0, T)$ and v(0) = 0. Then, $(u + \varepsilon v)$ is an admissible function and

$$J(u + \varepsilon v) = J(u) + 2\varepsilon \int_0^T (u'v' + \varphi(t)uv)\,dt \tag{34}$$

If J(u) is the minimum

$$\int_0^T (u'v' + \varphi(t)uv)\,dt = 0. \tag{35}$$

Integrating by parts the first term, we obtain

$$u'v\Big]_0^T + \int_0^T v[-u'' + \varphi(t)u]\,dt = 0. \tag{36}$$

Hence, we have the Euler equation

$$u'' - \varphi(t)u = 0 \tag{37}$$

with the condition u'(T) = 0.

This method assumes the existence of u". Following Haar we adopt another procedure and integrate the second term by parts and obtain

$$v\int_0^t \varphi(t_1)u(t_1)\,dt_1\Big]_0^T + \int_0^T \Big[u'v' - v'\int_0^t \varphi(t_1)u(t_1)\,dt_1\Big]dt = 0, \tag{38}$$

since v(0) = 0, we have

$$\int_0^T v'\Big[u' + \int_t^T \varphi(t_1)u(t_1)\,dt_1\Big]dt = 0 \tag{39}$$

for all $v' \varepsilon L^2(0, T)$. If we choose

$$v' = u' + \int_t^T \varphi(t_1)u(t_1)\,dt_1 \tag{40}$$

we find

$$\int_0^T \left[u' + \int_t^T \varphi(t_1)u(t_1)\right]^2 dt_1 = 0. \tag{41}$$

From (41) it follows that

$$u' + \int_t^T \varphi(t_1)u(t_1)\,dt_1 = 0. \tag{42}$$

Differentiating this, we obtain the Euler equation

$$u''(t) - \varphi(t)u(t) = 0 \tag{43}$$

with $u(0) = c$, and $u'(T) = 0$.

For the minimization procedure to hold the minimum requirement is that $J(u)$ should be positive definite for all nontrivial u. In some cases such as $\varphi(t) > 0$ this condition is obviously satisfied. In other cases, the condition holds depending on the magnitude of T. The integral $J(u)$ exists if $u' \varepsilon L(0, T)$ which means u is continuous. Let us assume that the Euler Equation (43) has a unique solution. To show this let us take the two fundamental solutions of (43) u_1, u_2, such that

$$u_1(0) = 1 \qquad u_1'(0) = 0 \tag{44}$$

$$u_2(0) = 0 \qquad u_2'(0) = 1. \tag{45}$$

Hence, any solution $u = cu_1 + bu_2$ where c and b are to be determined by the end conditions. The constant b can be uniquely determined if $u_2'(T) \neq 0$. This is secured by the assumption that $J(u)$ is positive definite.

Let us start with the relation

$$\int_0^T u_2(u_2'' - \varphi(t)u_2)\,dt = 0. \tag{46}$$

Integrating by parts

$$u_2u_2'\Big|_0^T - \int_0^T (u_2'^2 + \varphi(t)u_2^2)\,dt = 0. \tag{47}$$

If $u_2'(T)$ were to be zero, we have

$$\int_0^T (u_2'^2 + \varphi(t)u_2^2)\,dt = J(u_2) = 0. \tag{48}$$

This is a contradiction if $J(u)$ is positive definite since u_2 is not identically zero. Hence, there is a unique solution to Equation (43).

This solution also furnishes the minimum of $J(u)$. We see this easily since for any function v such that $v' \varepsilon L^2(0, T)$, $v(0) = 0$, we have

$$J(u + v) = J(u) + J(v) + 2\int_0^T (u'v' + \varphi(t)uv)\,dt. \tag{49}$$

The last term on the right-hand side of Equation (49), when integrated by parts, yields

$$\int_0^T (u'v' + \varphi(t)uv)\,dt = vu'\Big]_0^T - \int_0^T v(u'' - \varphi(t)u)\,dt = 0. \tag{50}$$

Since $u'(T) = 0$, we have

$$J(u + v) = J(u) + J(v) > J(u). \tag{51}$$

If there is any other function v which supposedly yields a value of J no larger than $J(u)$ that is $J(v) \leq J(u)$, then we see from the foregoing

$$J(v) = J(u + v - u) = J(u) + J(v - u) > J(u). \tag{52}$$

If v is not identically equal to u, this is a contradiction. Hence, we deduce the uniqueness of the function that minimizes $J(u)$. The positive definiteness of $J(u)$ regardless of the sign of $\varphi(t)$ for small values of T can also be established as in Reference [1].

Let us now introduce the basic idea of the Rayleigh-Ritz method to find the solution of the Euler Euqation (43).

$$u'' - \varphi(t)u = 0 \qquad u(0) = c \qquad u'(T) = 0. \tag{53}$$

Let us use the trial function

$$u = \sum_{k=1}^{N} a_k w_k \tag{54}$$

such that $u(0) = c = \Sigma_{k=1}^{N} a_k w_k(0)$. We might chose w_k in such way that this condition is automatically satisfied. For example, we may have

$$u = c + \sum_{k=1}^{N-1} a_k \sin kt. \tag{55}$$

Then

$$\begin{aligned} J(u) &= \int_0^T \left[\left(\sum_{k=1}^{N} a_k w_k'\right)^2 + \varphi(t)\left(\sum_{k=1}^{N} a_k w_k\right)^2\right] dt \\ &= Q(a_1, a_2, \cdots, a_N) \end{aligned} \tag{56}$$

a quadratic form in the coefficients $a_1, a_2, \ldots, a_N$. They are determined by minimizing Q with respect to each one of them. Thus, we have now the task of solving a system of linear algebraic equations.

The validity of this procedure and the monotone behavior and convergence of the coefficients have been demonstrated in the literature. We can choose N large enough to obtain an arbitrarily good approximation to the desired minimum value. The task of choosing the functions $\{w_k\}$ in some adroit fashion so as to obtain good approximations for moderate N is not an easy one. Physical intuition and a certain amount of experience play important roles.

Suppose we find an admissible function v such that

$$J(v) - J(u) \leq \varepsilon. \tag{57}$$

Then wat can be said about $|u - v|$? Let us recall the Euler equation associated with the minimization of J(u), which is

$$u'' - \varphi(t)u = 0 \qquad u(0) = c \qquad u'(T) = 0 \tag{58}$$

we have then

$$\int_0^T u(u'' - \varphi(t)u \, dt = 0. \tag{59}$$

Integration by parts yields

$$uu' \Big]_0^T - \int_0^T (u'^2 + \varphi(t)u^2) \, dt = 0. \tag{60}$$

Thus

$$J_{min} = - u'(0)c. \tag{61}$$

This yields the missing initial condition at $t = 0$ for the Euler equation. If we have found another approximate function v, and, hence, $J(v)$, so that we know

$$J(u) \leq J(v) \leq J(u) + \varepsilon. \tag{62}$$

We can use $-J(v)/c = v'(0)$ and $v(0) = c$, where v satisfies the Euler equation for v. This means

$$v'' - \varphi(t)v = 0 \qquad v(0) = c; \qquad v'(0) = -J(v)/c. \tag{63}$$

The actual u satisfies

$$u'' - \varphi(t)u = 0 \qquad u(0) = c; \qquad u'(0) = -J(u)/c. \tag{64}$$

If u_1 and u_2 are the fundamental solutions of the Euler equation, then

$$u = cu_1 - \frac{J(u)u_2}{c}; \qquad v = cu_1 - \frac{J(v)u_2}{c}. \tag{65}$$

Hence,

$$|u - v| = \left|\left\{\frac{J(u) - J(v)}{c}\right\}u_2\right| \quad \frac{k_2\varepsilon}{c} \tag{67}$$

$$\text{for} \quad 0 \leq t \leq T$$

The minimization of $J\left(\Sigma_{k=1}^{N} a_k w_k\right)$ leads to the problem of solving linear algebraic equations whose coefficients are determined by integrals of the form

$$\int_0^T w_k' w_\ell' \, dt \quad \text{and} \quad \int_0^T \varphi(t) w_k w_\ell \, dt. \tag{68}$$

We can choose the functions w_k's such that they satisfy either

$$\int_0^T w_k' w_\ell' \, dt = \delta_{k\ell} \tag{69}$$

or

$$\int_0^T \varphi(t) w_k w_\ell \, dt = \delta_{k,\ell} , \tag{70}$$

where $\delta_{k\ell}$ is the Kronecker delta. We cannot satisfy both the conditions simultaneously. These integrals can be evaluated by numerical quadratue. An approach which may be more convenient may be to solve for the set of equations

$$\frac{dr_{k\ell}}{dt} = w_k' w_\ell' \qquad r_{k\ell}(0) = 0 \tag{71}$$

$$\frac{dS_{k\ell}}{dt} = \varphi(t) w_k w_\ell \qquad S_{k\ell}(0) = 0. \tag{72}$$

The required integrals are $r_{k\ell}$ and $S_{k\ell}$. If function w_k satisfies a differential equation as is often the case, the foregoing procedure results in considerable decrease in the storage requirements of the computer.*

There is no difficulty in extending the foregoing considerations to the multidimensional case. Consider the linear differential equation

$$x'' = AX \qquad x(0) = c \quad \text{and} \quad x'(T) = 0 \tag{73}$$

---

* Either $r_{k\ell}$ or $S_{k\ell}$ can be chosen to be diagonal.

x is a function of n dimension and A is an n × n matrix. We associate with this Euler equation the functional

$$J(x) = \int_0^T [(x', x') + (x, Ax)]\, dt. \tag{74}$$

We can also adopt the same method of attack with nonlinear equations. Consider the equation

$$u'' - g(u) = 0; \quad u(0) = c, \text{ and } u'(T) = 0.$$

The functional to be minimized is

$$J(u) = \int_0^T [u'^2 + 2h(u)]\, dt \tag{75}$$

with $g(u) = \frac{dh(u)}{du}$ and $u(0) = c$.

In the Thomas-Fermi statistical model of a free neutral atom, we are to solve a nonlinear equation of the second order over the interval $(0, \infty)$ [4], [5]. The equation is

$$\varphi'' - x^{-1/2}\varphi^{3/2} = 0 \quad \varphi(0) = 1; \quad \varphi(\infty) = 0. \tag{76}$$

This is also known as the Emden-Fowler equation in astrophysics. We can see that this is the Euler equation - corresponding to the functional

$$J(\varphi) = \int_0^\infty (\varphi'^2 + \tfrac{4}{5}x^{-1/2}\varphi^{5/2})\, dx. \tag{77}$$

In this case the initial slope $\varphi'(0)$ is given by the integral

$$-\varphi'(0) = \int_0^\infty (\varphi'^2 + x^{-1/2}\varphi^{5/2})\, dx. \tag{78}$$

This is seen easily since

$$\int_0^\infty \varphi'^2\, dx = -\varphi'(0) - \int_0^\infty x^{-1/2}\varphi^{5/2}\, dx. \tag{79}$$

If we use for $\varphi(x)$ the function

$$\varphi(x) = (1 + x^{1/\alpha})^{-\beta} \quad (0 < x < \infty), \tag{80}$$

where α and β are the variational parameters, we can obtain for the min of $J(\varphi)$ and upper bound. For further details, the reader can refer to the literature cited.

The Rayleigh-Ritz method can also be applied to study linear integral equations of the type

$$u(t) + \int_0^a k(t, s)u(s)\, ds = f(t). \tag{81}$$

Suppose $k(t, s)$ is a real symmetric kernel, then (81) can be found to be the variational equation associated with the quadratic functional

$$J(u) = \int_0^a u^2(t)\, dt + \int_0^a \int_0^a k(t, s)u(t)u(s)\, dt\, ds -$$

$$- 2 \int_0^a f(t)u(t)\, dt. \tag{82}$$

The question of whether $J(u)$ has an absolute minimum leads to a characteristic value problem. Assuming that $J(u)$ does possess an absolute minimum, we can see easily that for the minimum value, $J(u)$ can be evaluated and we have

$$J_{min}(u) = - \int_0^a f(t)u(t)\, dt. \tag{83}$$

Hence, we can obtain approximate values for the quantity $\int_0^a f(t)u(t)\, dt$. For further details see Reference [22].

The Rayleigh-Ritz method may be made rigorous by using the Sturm-Liouville theory as is done in the next section.

4. STURM-LIOUVILLE THEORY

To indicate the connection between the Sturm-Liouville equations and the calculus of variations, we consider the homogeneous functional

$$K(u) = \frac{\int_0^T u'^2 \, dt}{\int_0^T \varphi(t) u^2(t) \, dt} \tag{84}$$

Let us determine u which renders k(u) stationary under the conditions

$$u(0) = u(T) = 0 \qquad u' \in L^2(0, T) \quad \text{and} \quad |\varphi| \in L(0, T). \tag{85}$$

Let $u = \bar{u} + \varepsilon v$, where $\bar{u}$ is in this class and v is an admissible function ε, a small scalar

$$v' \varepsilon L^2(0, T) \quad \text{and} \quad v(0) = v(T) = 0.$$

$$k(\bar{u} + \varepsilon v) = \frac{\int_0^T (\bar{u}' + \varepsilon v')^2 \, dt}{\int_0^T \varphi(t)(\bar{u} + \varepsilon v)^2 \, dt} \, . \tag{86}$$

Taking derivatives with respect to ε and equating it to zero, we obtain

$$\int_0^T \bar{u}' v' \, dt - \frac{\int_0^T \bar{u}'^2 \, dt}{\int_0^T \varphi(t) \bar{u}^2 \, dt} \int_0^T \varphi(t) \bar{u} v \, dt = 0 \tag{87}$$

for all admissible v. Introduce the parameter

$$\lambda = k(\bar{u}) = \frac{\int_0^T \bar{u}'^2 \, dt}{\int_0^T \varphi(t)\bar{u}^2 \, dt} . \tag{88}$$

Then we have

$$\int_0^T \bar{u}'v' \, dt - \lambda \int_0^T \varphi(t)\bar{u}v \, dt = 0 \tag{89}$$

valid for all admissible functions v. Integration by parts yields the Euler equation

$$\bar{u}'' + \lambda\varphi(t)\bar{u} = 0. \tag{90}$$

If we consider the initial value problem with $u(0) = 0$ and $u'(0) = 1$ (this assumption about $u'(0)$ being equal to 1 does not detract from the generality of the problem), it can be shown [1] that $u(t, \lambda)$ is an entire function of λ, for $0 \leq t \leq T$. Hence, the values of λ which provide nontrivial solutions of (90) satisfying the further condition $u(T) = 0$ are the zeroes of an entire function $u(T, \lambda)$. Thus we know that the admissible values of λ are denumerable and that the only cluster point is at ∞. We call these admissible values of λ characteristic values and the associated solutions eigenfunctions.

For the Sturm-Liouville euqation it is easy to establish the following basic results.

(1) The values of λ are real and positive and we can write $\lambda_1 < \lambda_2 < \lambda_3 \cdots$.

(2) The associated solutions $u_1, u_2, \cdots$ may be chosen real and they are orthogonal with the weight function $\varphi(t)$, i.e.,

$$\int_0^T \varphi(t)u_1u_j \, dt = 0 \quad i \neq j. \tag{91}$$

If further we assume $\varphi(t) > 0$, $u(T, \lambda) = 0$ cannot have a double root. From Equation (91) one can deduce by integration by parts and using Euqation (90)

$$\int u_i' u_j' \, dt = 0 \quad i \neq j. \tag{92}$$

We have used $u'(0) = 1$ called the normalization sometimes. However, we can use a more convenient normalization

$$\int_0^T \varphi u_i^2 \, dt = 1. \tag{93}$$

With this change we can write

$$\begin{aligned} \int_0^T \varphi(t) u_i u_j \, dt &= \delta_{ij} \\ \int_0^T u_i' u_j' \, dt &= \lambda_i \delta_{ij}. \end{aligned} \tag{94}$$

Let us now expand $u(t)$ in an orthogonal series using $u_k(t)$ themselves

$$u \sim \sum_{k=1}^{\infty} a_k \, u_k(t) \tag{95}$$

so that

$$a_k = \int_0^T \varphi u \, u_k \, dt. \tag{96}$$

Hence, by Parseval's relation we have

$$\int_0^T \varphi u^2 \, dt = \sum_{k=1}^{\infty} a_k^2. \tag{97}$$

Also, we have the result

$$\int_0^T u'^2 \, dt = \Sigma\lambda_k a_k^2 . \tag{98}$$

Now we can easily discuss the minimum value of

$$k(u) = \frac{\int_0^T u'^2 \, dt}{\int_0^T \varphi u^2 \, dt} \tag{99}$$

$$= \frac{\sum_{i=1}^{\infty} \lambda_i a_i^2}{\sum_{i=1}^{\infty} a_i^2} \tag{100}$$

for $u \varepsilon L^2(0, T)$ $u(0) = u(T) = 0$. Hence, from this representation the minimum value of $k(u)$ is given by

$$\min_u k(u) = \lambda_1, \tag{101}$$

where u runs over the admissible functions. Further the minimum is attained uniquely for the function u, which is proportional to u_1.

Furthermore, this representation clearly shows the significance of the higher characteristic values as relative minima. Thus

$$\lambda_2 = \min_{R_1} k(u), \tag{102}$$

where R_1 is the region of space spanned by functions u, such that

$$u(0) = u(T) = 0 \quad \text{and} \quad u'(t) \; \varepsilon \; L^2(0, T) \tag{103}$$

and

$$\int_0^T \varphi(t) u\, u_1\, dt = 0. \tag{104}$$

Similarly for the third root

$$\lambda_3 = \min_{R_2} k(u), \tag{105}$$

where R_2 is space determined by $u(0) = u(T) = 0$; $u'(T) \in L^2(0, T)$ and

$$\int \varphi(t) u u_1\, dt = 0 \tag{106}$$

$$\int \varphi(t) u u_2\, dt = 0 \tag{107}$$

and so on. For details see Courant and Hilbert, Reference [27].

Since in many physical cases the determination of the eigenvalues are important, a variational representation which provides a quick way to obtain the approximate values using simple algebraic procedures is the Rayleigh-Ritz method, which gives the lowest root as

$$\lambda_1 = \min_{\substack{u \\ u(0) = u(T) = 0}} \frac{\int_0^T u'^2\, dt}{\int_0^T \varphi(t) u^2\, dt} \tag{108}$$

The upper bounds for are obtained by choosing

$$u = \sum_{k=1}^{N} a_k v_k, \tag{109}$$

where $v_k(0) = v_k(T) = 0$, and v_k's are conveniently chosen functions, and a_k are scalars. We minimize the expression

$$\frac{\int_0^T \left(\sum_{k=1}^N a_k v_k'\right)^2 dt}{\int_0^T \varphi(t)\left(\sum_{k=1}^N a_k v_k\right)^2 dt} = \frac{(a,\ Aa)}{(a,\ Ba)}, \tag{110}$$

where a_k are the components of the vector a and the symmetric matrices A and B are given by

$$A_{k\ell} = \int_0^T v_k' v_\ell'\ dt, \tag{111}$$

$$B_{k\ell} = \int_0^T \varphi(t) v_k v_\ell\ dt. \tag{112}$$

This algebraic problem leads to the determinantal equation of order N namely $|A - \lambda B| = 0$. Not only does $\lambda_1^{(N)}$ give information concerning λ_1, but the other roots $\lambda_2^{(N)}, \lambda_3^{(N)}, \ldots, \lambda_N^{(N)}$ yield information concerning the eigenvalue in the sense that

$$\lambda_1^{(N)} \leq \lambda_2^{(N)} \leq \lambda_3^{(N)} \leq \cdots \lambda_N^{(N)}, \text{ etc.}$$

5. RAYLEIGH-RITZ METHOD AND PHYSICAL PROCESSES

The physical meaning of the method described above is clear when we consider the Rayleigh principle for a continuous system [7]. Consider the vibrations of a non-uniform string stretched under a tension T (constant). If the displacement of the string from equilibrium is y(x, t) at time t at a point x, it obeys the equation

$$\frac{d^2y}{dx^2} = \frac{\rho(x)}{T}\frac{d^2y}{dt^2} \tag{113}$$

where $\rho(x)$ is the mass per unit length. If it vibrates with frequency ω,

$$y(x, t) = u(x) \sin (\omega t + \varepsilon), \tag{114}$$

and the amplitude u(x) satisfies the Sturm-Liouville type equation

$$\frac{d^2u}{dx^2} + \frac{\omega^2\rho}{T} u = 0 \tag{115}$$

with $u(0) = u(\ell) = 0$ since the string is held fixed at its two ends $x = 0$ and $x = \ell$. The values ω which yield non-trivial solutions give the eigenfrequencies and the corresponding u(x) the eigen-solutions. Oscillating with the proper eigen-values of the system, the frequencies are to be found by the principle that the average potential energy over the string should equal the average kinetic energy.

We suppose that v(x) is an arbitrary function satisfying the end conditions $v(0) = v(\ell) = 0$ and if the string vibrates with frequency ℓ and amplitude v(x) the mean kinetic and potential energies are given by

$$\overline{K \cdot E} = \frac{1}{4} \omega^2 \int_0^\ell \rho(x)[v(x)]^2 \, dx \tag{116}$$

and

$$\overline{Pot \cdot E} = \frac{1}{4} T \int_0^\ell \left(\frac{dv}{dx}\right)^2 dx. \tag{117}$$

Then ω^2 should be obtained from

$$\omega^2 = \frac{T \int_0^\ell \left(\frac{dv}{dx}\right)^2 dx}{\int_0^\ell \rho(x)\{v(x)\}^2 \, dx} \tag{118}$$

Equation (118) is the same type of functional that we used for the variational calculations earlier.

If $u_1(x)$ denotes the solution of (115) for the lowest frequency ω_1, then we have

$$\frac{\omega_1^2}{T}\,\rho(x) = -\frac{d^2u_1}{dx^2}/u_1. \tag{119}$$

Hence,

$$\left(\frac{dv}{dx}\right)^2 - \frac{\omega_1^2\rho v^2}{T} = \left(\frac{dv}{dx}\right)^2 + \left(\frac{d^2u_1}{dx^2}\right)\cdot\frac{1}{u_1} \tag{120}$$

$$= \frac{d}{dx}\left(\frac{\frac{du_1}{dx}v^2}{u_1}\right) + \left(\frac{\frac{du_1}{dx}v - u_1\frac{dv}{dx}}{u_1}\right)^2 \tag{121}$$

Therefore, we have the result

$$\int_0^\ell \left(\frac{dv}{dx}\right)^2 dx - \frac{\omega_1^2}{T}\int_0^\ell \rho(x)[v(x)]^2\,dx \geq \left[\left(\frac{\frac{du_1}{dx}v^2}{u_1}\right)\right]_0^\ell. \tag{122}$$

Because of the boundary conditions on v, the right-hand side of (122) vanishes and we are led to conclude that

$$\int_0^\ell \left(\frac{dx}{dx}\right)^2 dx \geq \frac{\omega_1^2}{T}\int_0^\ell \rho(x)[v(x)]^2\,dx. \tag{123}$$

That is

$$\omega^2 \geq \omega_1^2. \tag{124}$$

The equality obtains only if $v = u$. Thus the frequencies obtained by equating the mean values of the kinetic and potential energies of the system satisfying the appropriate boundary conditions provides an upper bound to the lowest eigenvalues of the system.

For a general problem described by the Sturm-Liouville equation

$$\frac{d}{dx}\left\{p(x)\,\frac{du}{dx}\right\} + \left\{q(x) + \lambda r(x)\right\}u(x) = 0, \tag{124}$$

where λ is a constant, and p(x) has no zeroes in the range 0 to T and r(x) remains positive within the range of definition, the functional to be minimized as constructed earlier is given by

$$J(u) = \frac{\int_0^T \left\{p(x)\left(\frac{du}{dx}\right)^2 - q(x)u^2\right\} dx}{\int_0^T r(x)[u(x)]^2 \, dx} \qquad (125)$$

By the Rayleigh-Ritz procedure again we end up with the determinantal equation of the type $|A - \lambda B| = 0$.

For the three dimensional Helmholtz equation of the form

$$\nabla^2\Psi + \frac{\omega^2}{c^2}\Psi = 0 \qquad (126)$$

we start multiplying by Ψ and integrating over the spatial coordinate. We have

$$\frac{\omega^2}{c} = \frac{\int_\Omega \Psi\nabla^2\Psi d\Omega}{\int_\Omega \Psi^2 d\Omega} . \qquad (127)$$

Now suppose that at the boundary surface S of the volume of integration we have either

$$\text{(i)} \quad \Psi = 0 \quad \text{or}$$

$$\text{(ii)} \quad \frac{\partial\Psi}{\partial\vec{n}} = 0, \qquad (128)$$

where $\vec{n}$ is the unit vector in the direction of the outward drawn normal to S. Then by Green's theorem we see that

$$\int \Psi\nabla^2\Psi \, d\Omega + \int (\nabla\Psi)^2 \, d\Omega = \int_S \Psi \frac{\partial\Psi}{\partial n} dS = 0 \qquad (129)$$

and so it follows

$$\frac{\omega^2}{c^2} \quad \frac{\int_\Omega (\nabla\Psi)^2 \, d\Omega}{\int_\Omega \Psi^2 \, d\Omega} \tag{130}$$

and the functional $J(\chi)$

$$J(\chi) = \frac{\int_\Omega (\nabla\chi)^2 \, d\Omega}{\int_\Omega \chi^2 \, d\Omega} \quad \text{and} \quad J(\chi) \geq 0, \tag{131}$$

when minimized over admissible χ functions will give an upper bound for ω^2/c^2. To see that $J(\Psi)$ is stationary, consider the function $\Psi + \delta\Psi$ a slight variation from Ψ with the same boundary conditions (i) or (ii) on the surface S. Then to first order in small quantities we have

$$\delta J \int \Psi^2 \, d\Omega + 2J \int (\delta\Psi)\Psi \, d\Omega = 2 \int \nabla(\delta\Psi) \cdot (\nabla\Psi) \, d\Omega. \tag{132}$$

By Green's theorem we have

$$\int (\nabla\delta\Psi) \cdot (\nabla\Psi) \, d\Omega + \int \delta\Psi\nabla^2\Psi d\Omega - \int \delta\Psi \, \frac{\partial\Psi}{\partial\vec{n}} \, dS. \tag{133}$$

The right-hand side vanishes since $\delta\Psi$ or $\partial\Psi/\partial n$ is zero on S. Since Ψ satisfies the Helmholtz equation, we have

$$\delta J(\Psi) = \frac{2\int \delta \, \Psi \, (\nabla^2\Psi + \frac{\omega^2}{c^2} \Psi) \, d\Omega}{\int \Psi^2 \, d\Omega} = 0. \tag{134}$$

Therefore $J(\Psi)$ is stationary. Conversely, the function Ψ which yields stationary values of $J(\Psi)$ satisfies the Helmholtz equation. This concept leads to a method which had been successfully applied to quantum mechanical systems described by the Schrödinger equation in connection with determination of their eigen frequencies.

$$(H - E)\Psi = 0, \tag{135}$$

where H is Hermitian.

The functional to be minimized is given by

$$J(\chi)= \frac{\int \bar{\chi} H \chi \, d\Omega}{\int \bar{\chi}\chi \, d\Omega} \tag{136}$$

and for the correct solution Ψ we get $J(\Psi) = E$ (χ attains a minimum value Ψ). The stationary nature of this functional is easily shown employing suitable boundary conditions [7]. The calculation of the upper bounds to the various eigen-energies can be done as in the one-dimensional case. This was first applied by Hylleraas-Undheim [8] to calculate the excited state energies of atoms. The variational method has become a powerful tool in calculations concerning many electron atoms and molecules, in polarizability calculation and in many other fields in theoretical physics.

In scattering problems, we are interested in the phase shifts. We can obtain a variational formulation for the calculation of phase shifts. We start with the radial Schrödinger equation of the ℓ-th partial wave

$$u_\ell'' + \left[k^2 - v(r) - \frac{\ell(\ell + 1)}{r^2}\right]u_\ell = 0$$

or (137)

$$Lu_\ell = F(r)$$

with

$$L = \left[\frac{\partial^2}{\partial r^2} + k^2 - \frac{\ell(\ell + 1)}{r^2}\right]$$

(138)

$$F(r) = v(r)u_\ell .$$

The usual boundary conditions are that $u_\ell(r)$ is regular at the origin and goes as $\sin\left(kr - \ell\frac{\pi}{2} + \delta\right)$ at ∞. The homogeneous equation

$$Lu_\ell = 0 \tag{139}$$

has the fundamental solutions $j_\ell(kr)$ and $n_\ell(kr)$ which behave at ∞ as

$$j_\ell(kr) \to \sin\left(kr - \ell\frac{\pi}{2}\right), \tag{140}$$

$$n_\ell(kr) \to -\cos(kr - \frac{\ell\pi}{2}). \tag{141}$$

The solution of Equation (137) can be written as

$$u_\ell(r) = j_\ell(kr) + \int_0^\infty dr' G(r, r') v(r') u_\ell(r'), \tag{142}$$

where the Green's function G(r, r') is given by

$$G(r, r') = \begin{cases} \frac{1}{k} j_\ell(kr) n_\ell(kr') & r < r' \\ \\ \frac{1}{k} n_\ell(kr) j_\ell(kr') & r > r'. \end{cases} \tag{143}$$

Let us make the kernel of the integral equation symmetric by the following substitutions

$$\begin{aligned} y(r) &= \sqrt{v(r)}u_\ell(r); \qquad f(r) = \sqrt{v(r)}j_\ell(kr) \\ k(r, r') &= \sqrt{v(r)v(r')}G(r, r') = k(r', r), \end{aligned} \tag{144}$$

with these changes, the integral equation (142) becomes

$$y(r) = f(r) + \int_0^\infty dr' k(r, r') y(r') \, dr' \tag{145}$$

with the symmetric kernel

$$k(r, r') = k(r', r). \tag{146}$$

The solution of the integral equation (142) for large values of r can be written as

$$u_\ell(r) \to \sin(kr - \frac{\ell\pi}{2}) - \frac{1}{k}\cos(kr - \frac{\ell\pi}{2}) \int_0^\infty dr' v(r') j_\ell(kr') u_\ell(r'). \tag{147}$$

Hence, the phase shift is given as

$$\tan \delta_\ell = -\frac{1}{k}\int_0^\infty dr\ v(r) j_\ell(kr) u_\ell(r). \tag{148}$$

By a reference to Equation (144) the integral

$$I_2 = \int_0^\infty dr\ y(r) f(r) = -\ k \tan \delta_\ell. \tag{149}$$

From Equation (145) we can easily see that if we form the integral I_1

$$I_1 = \int dr\ y(r) \left\{ y(r) - \int_0^\infty dr' k(r, r') y(r') \right\}, \tag{150}$$

I_1 and I_2 are the same for the correct solution y of the equation.

Since k(r, r') is symmetric and since y is the solution of the integral Equation (145), we find that $\delta I_1 = 2\delta I_2$ only for the correct solution y. This means

$$\delta\left(\frac{I_1}{I_2^2}\right) = 0 \tag{151}$$

for the exact solution and provides a method of determining I_1/I_2^2 which for the extremum value will turn out to be

$$S_\ell = -\frac{1}{k}\cot\delta_\ell = \frac{\int dr\ v(r) u_\ell(r) \left\{ u_\ell(r) - \int_0^\infty dr' v(r') u_\ell(r') G(r, r') \right\}}{\left[\int_0^\infty dr\ v(r) u_\ell(r) J_\ell(kr) \right]^2}. \tag{152}$$

The extremal value of S_ℓ determines the correct phase shift. This could also have been obtained from Equation (82) relating to integral equations with symmetric kernels. Equation (152) is known as Schwinger's variational principle (see Reference [20]) for the phase shift. Similar variational formulations for the phase shifts have been given by others also, see References [6, 18, 19], and [28a, b].

6. THE MAXIMUM FUNCTIONAL

In these variational methods we have obtained the upper bounds by minimizing a functional of the type

$$J(u) = \int_0^T (u'^2 + 2g(u))\, dt \tag{153}$$

subject to the condition $u(0) = c$, and $u'(T) = 0$. For any admissible function v, it has been shown that

$$\min_u J(u) \leq J(v). \tag{154}$$

The value of these methods will be tremendously enhanced if we can devise some method by which we can obtain lower bounds also. This can be done by using the concept of duality. We will present some simple results using the theme of duality dealt with in References [10], [11], and [12].

To present the basic formalism, we start with the observation

$$u'^2 = \max_v [2u'v - v^2]. \tag{155}$$

We have used this idea when we dealt with quasilinearization methods in Chapter VI. Using Equation (155), we can write

$$\int_0^T [u'^2 + 2g(u)]\, dt = \max_v \left\{ \int_0^T [2u'v - v^2 + 2g(u)]\, dt \right\}. \tag{156}$$

Hence, for any function $v \in L^2(0, T)$ we have

$$\int_0^T [u'^2 + 2g(u)]\, dt \geq \left\{ \int_0^T [2u'v - v^2 + 2g(u)]\, dt \right\}. \tag{157}$$

Hence,

$$\min_u \int_0^T [u'^2 + 2g(u)]\, dt \geq \min_u \left\{ \int_0^T [2u'v - v^2 + 2g(u)]\, dt \right\}. \tag{158}$$

We further restrict the class of v such that v' exists.

Since the equality in (155) is obtained for $v = u'$, we impose another restriction that $v(T) = 0$. Since we are interested only in obtaining bounds, we can choose v to our own satisfaction. With these restrictions we see that

$$\int 2u'v\,dt = 2uv\Big]_0^T - \int_0^T 2uv'\,dt \tag{159}$$

$$= -2cv(0) - \int_0^T 2uv'\,dt. \tag{160}$$

We therefore see that the right-hand side of Equation (158) becomes

$$\min_u\left[-2cv(0) + \int_0^T (2g(u) - 2uv')\,dt - \int_0^T v'^2\,dt\right]. \tag{161}$$

Let us now introduce a transform defined by

$$h(w) = \min_u (g(u) - wu), \tag{162}$$

$h(w)$ is called the Legendre-Fenchel transform [14], [15]. If g is a strictly convex function, i.e., if $g'' > 0$ for $-\infty < u < \infty$ then a unique minimizing value exists given by the solution of the equation

$$g_u - w = 0 \tag{163}$$

and u can be obtained as a function of w. For interesting properties of this transform, see References [1] and [17] cited. Hence, carrying out the minimizations with respect to u, we obtain from Equation (161)

$$\min_u J(u) \geq \left[-2cv(0) + \int_0^T (2h(v') - v^2)\,dt\right] \tag{164}$$

for all v satisfying the constraint

$$v(T) = 0. \tag{165}$$

Also v is such that $\int_0^T h(v')\,dt$ exists. Hence, we obtain a lower bound for $\min_u J(u)$ by maximizing the functional

$$k(v) = -2cv(0) + \int_0^T (2h(v') - v^2)\, dt. \tag{166}$$

This means

$$\min_u J(u) \geq \max_v k(v), \tag{167}$$

where v is taken over the functions of the foregoing nature. Under the assumption of convexity of g(u), we actually have equality.

$$\min_u J(u) = \max_v k(v) \tag{168}$$

when u and v have the end conditions u(0) = c, v(T) = 0, and are restricted to the appropriate functionsclasses. Compare this with Equation (154).

To realize the way in which the equality in (168) comes about, let us take g(u) to be a simple quadratic, say $u^2/2$. Then, we have

$$J(u) = \int_0^T (u^2 + u^2)\, dt \geq \max_v \left[-2cv(0) - \int_0^T (v'^2 + v^2)\, dt\right] \tag{169}$$

since h(v') becomes v'^2 in this case.

The Euler equation associated with the variational functional k(v) turns out to be

$$v'' - v = 0; \quad v'(0) = c; \quad v(T) = 0. \tag{170}$$

The Euler Equation related to J(u) as already found is

$$u'' - u = 0; \quad u(0) = c; \quad u'(T) = 0. \tag{171}$$

Comparing these two we can easily observe that

$$v = u'. \tag{172}$$

One can check that this satisfies the differential equations (170) and (171) along with the end conditions. The uniqueness of the solution of (170) can be easily established and hence we can use the identity of v with u' to obtain the maximum value of k(v). From Equation (170), we have

$$\int_0^T v(v'' - v)\,dt = 0; \qquad vv'\Big]_0^T - \int_0^T (v'^2 + v^2)\,dt = 0$$

$$- v(0)c - \int_0^T (v'^2 + v^2)\,dt = 0. \tag{173}$$

Thus for the maximum value of k(v) which we obtain for the solution which satisfies the Euler equation, we have

$$\max_v \left[-2cv(0) - \int_0^T (v'^2 + v^2)\,dt\right] = -cv(0) = -cu'(0). \tag{174}$$

We know from the earlier Equation (61) that

$$\min_u J(u) = -cu'(0) \tag{175}$$

which establishes the equality.

Thus when we wish to obtain numerical results without treating Euler equations, we may use trial functions x(t) and y(t) given by a series of suitable functions with coefficients which can be determined by the minimization procedure in one case and the maximization procedure in the case of the dual functions. This will lead us to obtaining bounds for the initial slope u'(0) the missing initial condition for the original equation we started to solve. Sometimes it may be advantageous to use for the trial function for minimizing J(u) a series

$$u = \sum_{k=1}^{N} a_k w_k \tag{176}$$

obeying the end conditions and take for v the trial function

$$v = \sum_{k=1}^{N} a_k w_k' \tag{177}$$

since we know that u' = v. For the multidimensional analog, we take

$$J(\vec{x}) = \int_0^T \left[(\vec{x}', \vec{x}') + (\vec{x}, A\vec{x}) \right] dt \tag{178}$$

and $\vec{x}(0) = \vec{c}$ where $\vec{x}$ is a vector and the A the corresponding matrix occurring in the Euler equation. Also, $\vec{x}'(T) = 0$; $A > 0$. The dual function k(y) can be found to be

$$k(\vec{y}) = \left[-2(\vec{c}, \vec{y}(0)) - \int_0^T (\vec{y}', A^{-1}\vec{y}') + (\vec{y}, \vec{y})\, dt \right] \tag{179}$$

with $\vec{y}(T) = 0$.

We will also find that

$$\min_{\vec{x}} J(\vec{x}) \geq \max_{\vec{y}} k(\vec{y}). \tag{180}$$

7. DYNAMIC PROGRAMMING METHOD

We will now apply the basic ideas of dynamic programming to obtain the minimum value of the functional J(u) which is necessary to the study of the solutions of the associated Euler equations as we have seen in earlier sections. In using dynamic programming techniques, we minimize the chosen variables over a number of sequential stages. In doing so we use the 'principle of optimality' which is the corner stone of dynamic programming. It states that

> "An optimal policy has the property that whatever the initial state and initial decisions are the remaining decisions must constitute an optimal policy for the state resulting from the first decision".

Utilization of this principle guarantees that the decision made at each stage is the best decision over the entire process. Also, we imbed the process in a larger class of similar processes which permits us to study a larger class of problems. We are also enabled to study the structure and sensitivity of the solution.

Consider the problem of minimizing

$$J(u) = \int_0^T g(u, u')\, dt \tag{181}$$

over all functions u(t) satisfying u(0) = c. Let us write this as a function of two variables

$$f(c, T) = \min_{u} \int_{0}^{T} g(u, u')\, dt. \tag{182}$$

Thereby we are imbedding the problem in a family of problems in which c and T are parameters. c can assume any value, while T > 0. The problem of determining a function u which minimizes J(u) may be considered as a problem of finding u', which minimizes J. We describe u' as a function of u and t, at each point so that we have a multistage process of the continuous type, if the u' at each point is the chosen policy.

A choice of u' over the interval [0, T] consists of a choice over [0, S] plus a choice over [S, T]. As a result of the choice of u' over [0, S] the initial value c is transformed into

$$\bar{c} = c + \int_{0}^{S} u'(t)\, dt. \tag{183}$$

From the additive character of the integral and the principle of optimality, we deduce the functional relation

$$f(c, T) = \min_{u'(0),\, S} \left[\int_{0}^{S} g(u, u')\, dt + f\left(c + \int_{0}^{S} u'(t)\, dt,\ T - S\right)\right]. \tag{184}$$

Without too much attention to rigor, we proceed in a formal fashion and consider S to be infinitesimal. The choice of u' over this interval comes down to the choice of u' at t = 0,

$$v = v(c, T) = u'(0), \tag{185}$$

indicating that the optimal slope is a function of the initial state c, and the duration of the process T. Expanding the quantities in (184) around S = 0, we have

$$\int_{0}^{S} g(u, u')\, dt = g(c, v)S + 0(S^2) \tag{186}$$

$$c + \int_0^S u'(t)\, dt = c + vS + 0(S^2) \tag{187}$$

$$f\left[c + \int_0^S u'\, dt,\ T - S\right] = f[c,\ T] + Sv\,\frac{\partial f}{\partial c} - S\,\frac{\partial f}{\partial T} + 0(S^2). \tag{188}$$

Hence, we easily see that in the limit $S \to 0$ (see Equations (184) and (186))

$$\frac{\partial f}{\partial T} = \min_v [g(v,\ c) + v\,\frac{\partial f}{\partial c}]. \tag{189}$$

The initial condition is clearly

$$f(c,\ 0) = 0. \tag{190}$$

Of course, the minimization of v in (189) gives another equation

$$\frac{\partial g}{\partial v} + \frac{\partial f}{\partial c} = 0. \tag{191}$$

Hence, using this value of v which minimizes the bracket in (189), we have

$$\frac{\partial f}{\partial T} = g - vg_v \tag{192}$$

and

$$\frac{\partial f}{\partial c} = -g_v. \tag{193}$$

From these two equations one can derive the usual Euler equation by different methods. Suppose we call

$$f(c,\ t) = \min_u J(u) = \min_u \int_0^T g(u,\ u')\, dt \tag{194}$$

we will arive at

$$-\frac{\partial f}{\partial t} = \min_v\ [g(c,\ v,\ t) + \frac{\partial f}{\partial c}]. \tag{195}$$

Let us call v, u' and c, u and obtain the set of coupled equations

$$g_{u'} + f_u = 0 \tag{196}$$

$$g + u'f_u = -f_t. \tag{197}$$

Differentiating (196) with respect to t, and (197) with respect to u, we obtain the relations

$$\frac{d}{dt}(g_{u'}) + f_{ut} + f_{uu}u' = 0 \tag{198}$$

$$g_u + u'f_{uu} + f_{tu} = 0. \tag{199}$$

Equating f_{tu} and f_{ut} from these equations, we obtain the Euler equation

$$\frac{d}{dt}(g_{u'}) = g_u. \tag{200}$$

As an illustration we can consider the quadratic functional

$$J(u) = \int_0^T (u'^2 + b(t)u^2)\, dt. \tag{201}$$

We minimize this subject to the condition u(0) = c. To take care of the t dependence of the integrand, let us slightly modify this and obtain

$$f(c, a) = \min_u J(u) = \int_a^T (u'^2 + b(t)u^2)\, dt. \tag{202}$$

Here $0 \leq a \leq T$ and the initial condition is u(a) = c. Using (195), we obtain

$$-\frac{\partial f}{\partial a} = \min_v [v^2 + b(a)c^2 + v\frac{\partial f}{\partial c}] . \tag{203}$$

This leads to

$$-\frac{\partial f}{\partial a} = b(a)c^2 - \frac{1}{4}(\frac{\partial f}{\partial c})^2 \tag{204}$$

with the initial condition

$$f(c, T) = 0. \tag{205}$$

This is a nonlinear partial differential equation, but we can reduce this to an ordinary equation of the Riccati type in view of the results obtained in earlier sections (Equation (61)).

$$f(c, a) = r(a)c^2, \tag{206}$$

where r(a) is a function only of a.

Hence, Equation (204) becomes

$$-r'(a) = b(a) - r^2(a) \qquad r(T) = 0 \tag{207}$$

which is a simple initial value problem. Since $-\frac{1}{2}\frac{\partial f}{\partial c}(c, a) = v(a)$, we also know the value of u' at all points in the interval [0, T]. Thus the solution of the Euler equation with boundary conditions u(0) = c and u'(T) = 0 is completely known.

A multidimensional version of the above example, where dynamic programming can be applied quite easily corresponds to the matrix equation of the form

$$\vec{x}' = B\vec{x} \ ; \quad \vec{x}(0) = \vec{c} \tag{208}$$

and the functional to be minimized can be found as

$$J(\vec{x}) = \int_0^T [(\vec{x}', \vec{x}') + (\vec{x}, B(t)\vec{x})] \, dt. \tag{209}$$

As found in the scalar case, we can arrive at the equation for f(c, a) setting $\vec{x}(a) = \vec{c}$

$$f(c, a) = \min_{\vec{x}} J(\vec{x}) = \int_a^T [(\vec{x}', \vec{x}') + (\vec{x}, B(t)\vec{x})] \, dt \tag{210}$$

we have

$$-\frac{\partial f}{\partial a} = \min_{\vec{v}} [(\vec{v}, \vec{v}) + (\vec{c}, B\vec{c}) + (\text{grad } f, \vec{v})]. \tag{211}$$

The minimizing $\vec{v}$ is given by

$$\vec{v} = -\frac{\text{grad } f}{2}. \tag{212}$$

This reduces Equation (211) to the nonlinear equation

$$-\frac{\partial f}{\partial a} = (\vec{c}, B(a)\vec{c}) - \left(\frac{\text{grad } f, \text{ grad } f}{4}\right) \tag{213}$$

with

$$f(\vec{c}, T) = 0. \tag{214}$$

Since we know that

$$f(\vec{c}, a) = (\vec{c}, R(a)\vec{c}), \tag{215}$$

we see that R(a) satisfies the Riccati equation

$$-R'(a) = B(a) - R^2(a); \qquad R(T) = 0 \tag{216}$$

which can be solved as an initial value problem. The solution of these initial value problems, both in the scalar and the matrix case, are much more stable than solving the corresponding Euler equations with boundary values.

In the numerical calculations, the problem of minimization and the solution of the minimum functional is done by writing the discrete analog of Equation (184) for small S, i.e., S = Δ. Let us write down

$$f(\bar{u}, t) = \min_u \int_t^T g(u, u')\, dt \qquad 0 \le t \le T \tag{217}$$

with $x(t) = \bar{u}$ in the discrete form, by writing $t = n\Delta$. Such n runs from 0 to N.

Then the discrete version of our dynamic programming equation becomes

$$f_n(\bar{u}) = \min_w [g_n(\bar{u}, w)\Delta + f_{n+1}(\bar{u} + w\Delta)], \tag{218}$$

where $w = u'$. If $g(u, u')$ is of the form

$$g(u, u') = u'^2 + bu^2 \tag{219}$$

$$f_n(u_n) = \min_{w_n}\left[\left(b_n u_n^2 + w_n^2\right) + f_{n+1}(u_n + w_n)\right]. \tag{220}$$

The initial condition at n = N will be for free end conditions

$$f(u, T) = 0 = f_N(u) \tag{221}$$

and we iterate Equation (220) backwards until we come to $f_0(c)$. Any other end condition at t = T can be suitably introduced when $f_{N-1}(u)$ is calculated. Also one notices that $f_n(u_n)$ will be of the form

$$f_n(u_n) = c_n u_n^2. \tag{222}$$

Hence, Equation (220) may be rewritten as

$$f_n(u_n) = c_n u_n^2 = \min_{w_n}\left[\left(b_n u_n^2 + w_n^2\right)\Delta + c_{n+1}(u_n + w_n \Delta)^2\right]. \tag{223}$$

Recursion relations can be easily established for c_n's, with $c_N = 0$. The reader can easily work out these relations for himself. The characteristic value problem of determining the solutions of the equation

$$u'' + \lambda\phi(t)u = 0 \quad \text{with} \quad u(0) = u(1) = 0 \tag{224}$$

under reasonable assumptions on $\phi(t)$ can be considered as determining the relative minima for the functional

$$J(u) = \int_0^1 u'^2 \, dt \tag{225}$$

subject to the constraints

$$\int_0^1 \phi(t)u^2 \, dt = 1 \tag{226}$$

and

$$u(0) = u(1) = 0. \tag{227}$$

To find the absolute minimum of (225) is to have a discrete version of the variational question by writing the quadratic form

$$Q(x_1, x_2, x_3, \ldots, x_N) = (x_1 - c)^2 + (x_2 - x_1)^2 + \cdots (x_N - x_{N-1})^2 + x_N^2 \tag{228}$$

subject to the constraint

$$\sum_{i=1}^{N} \phi_i x_i^2 = 1. \tag{229}$$

Let us suppose that $0 < a < \phi_i < b < \infty$ for all i and introduce the sequences $f_k(c)$, $k = 1, 2, \ldots, N-1$ defined by the relation

$$f_k(c) = \min_x\{(x_k - c)^2 + (x_{k+1} - x_k)^2 + (x_{k+2} - x_{k+1})^2 + + \cdots x_N^2\}. \tag{230}$$

where x_k's are subject to the constraint

$$\sum_{i=k}^{N} \phi_k x_k^2 = 1. \tag{231}$$

It is easily seen that

$$f_{N-1}(c) = \min_{x_{N-1} x_N} [(x_{N-1} - c)^2 + x_N^2], \tag{232}$$

where

$$\phi_{N-1} x_{N-1}^2 + \phi_N x_N^2 = 1 \tag{233}$$

and hence we arrive at

$$f_k(c) = \min_{x_k}\left[(x_k - c)^2 + (1 - \phi_k x_k^2)^{1/2} f_{k+1}\left(\frac{x_k}{\left[1 - \phi_k x_k^2\right]^{\frac{1}{2}}}\right)\right], \tag{234}$$

where $\phi_k x_k^2 \leq 1$.

The value of $f_1(0)$ is an approximation to the lowest characteristic value of the Sturm-Liouville equation.

For the higher order equations, the dynamic programming approach may prove very useful if the imbedding is done in ingenious ways. For example, if we want to attack the partial differential equation

$$u_{xx} + u_{yy} = 0 \qquad (x, y) \in R \tag{235}$$

with u specified on the boundary of R, we can discretize the problem and replace (235) by

$$\frac{u(x + \Delta, y) + u(x - \Delta, y) - 2u(x, y)}{\Delta^2} + \frac{u(x, y + \Delta) + u(x, y - \Delta) - 2u(x, y)}{\Delta^2} = 0 \tag{236}$$

This may be rewritten as

$$u(x, y) = \frac{u(x + \Delta, y) + u(x - \Delta, y) + u(x; y + \Delta) + u(x, y - \Delta)}{4} \tag{237}$$

expressing the fact that u(x, y) is the mean of the values of u at the four nearest neighbors at (x, y). To determine the values of u inside a given region when the values of the boundary are given using (237) leads to a system of algebraic equations of large dimensions. To simplify the task we apply another type of approach based on dynamic programming and consider the functional

$$J(u) = \min_{u(x, y)} \int_R (u_x^2 + u_y^2)\, dx\, dy \tag{238}$$

whose Euler equation is (235) and replace the derivatives by

$$u_x = \frac{u(x + \Delta, y) + u(x, y)}{\Delta} \tag{239}$$

$$u_y = \frac{u(x, y + \Delta) + u(x, y)}{\Delta} . \tag{240}$$

For simplicity we consider R to be a rectangular region of length a and width b, with $a = N\Delta$ and $b = M\Delta$. Let a be the extension along the x direction and b the extension along the y direction. We consider the values of a vector R_i whose components are the values of u(x, y) at the M grid points at each of the ordinate values $x = 0, \Delta, 2\Delta, 3\Delta$, etc. That is, $R_0 = u(0, 0), u(0, \Delta), u(0, 2\Delta), \cdots, u(0, M\Delta)$, the next adjacent vector is $\vec{R}_1 = u(\Delta, 0), u(\Delta, \Delta), u(\Delta, 2\Delta), \cdots, u(\Delta, M\Delta)$; and so on, for R_3, R_4, ..., up to R_N. Using the variational principle, we can convert the minimization problem of Equation (238) into a multistage decision process for determining $\vec{R}_i$ the vector along each successive ordinate. This can be handled by dynamic programming. Other ingenious devices for irregular regions and higher dimensional equations are explained in the book by Angel and Bellman [21].

Let us make the following remarks. Dynamic programming may be regarded as invariant imbedding plus optimization.

Using the calculus of variations, in the classical method one seeks a curve $u = u(t)$ defined over a given interval $[0, T]$ which maximizes. The unknown function u is considered as a point in function space. In the present approach we have a continuous decision process, and at each point we seek a direction which is optimal, the solution is obtained in the form of the policy, a set of instructions for carrying out the process, i.e., we determine the slope of the curve at each point. In geometric parlance we can say that the classical view is that of a curve as a locus of points while dynamic programming considers a curve to be the envelope of tangents. Hence the two theories are dual to each other, a fact which manifests itself constantly. It follows that the use of the two approaches will be more powerful and more enlightening than the use of one or the other method alone. This holds true for both analytic and computational aspects.

One of the advantages of the dynamic programming approach is that it allows the treatment of constraints in a very easy fashion.

REFERENCES

[1] Bellman, R., Methods of Nonlinear Analysis, Vol. I, Ch. 5, Academic Press, Inc., New York, 1970.

[2] Finlayson, B.A. and L.E. Scriven, 'The Method of Weighted Residuals; A Review', Applied Math. Review 19 (1966), 735.

[3] Urabe, M., 'Galerkins' Procedure for Nonlinear Periodic Systems and Its Extensions to Multipoint Boundary Value Problems for General Nonlinear Systems', Numerical Solutions of Nonlinear Differential Equations, Wiley & Sons, Inc., New York, 1966.

[4] Ikebe, J. and T. Kato, 'Application of Variational Method to Thomas Fermi Equation', Journal Phys. Soc. Japan 12 (1957), 201.

[5] Atepanova, T.V., 'Generalized Emden-Fowler Equation', Differencial' nye Uravenenija 3 (1967), 831.
[6] Schwinger, J., Phys. Rev. 72 (1947), 742A.
[7] Moiseiwitsch, B.L., 'Variational Principles', Interscience Publications, New York, 1966.
[8] Hylleraas, E.A. and B. Undheim, Physik 65 (1930), 759.
[9] Bellman, R., 'Quasilinearization and Upper and Lower Bounds for Variational Problems', Quart. Applied Math. 19 (1962), 349.
[10] Beckenbach, E.F. and R. Bellman, Inequalities, Springer-Verlag, Berlin, 1965.
[11] Rockefeller, R.T., Convexity and Duality, Princeton University Press, New Jersey, 1968.
[12] Mond, B. and M.A. Hanson, 'Duality for Variational Problems', Journal of Mathematical Analysis and Applications 18 (1967), 355.
[13] Dreyfus, S. and M. Fiemer, 'A Ne Approach to the Duality Theory of Mathematical Programming', Rand Report No. P-2334, 1961.
[14] Bellman, R., Introduction to the Mathematical Theory of Control Porcesses, Vol. II, Ch. 9, Academic Press, Inc., New York, 1971.
[15] Bellman, R., Stability Theory of Differential Equations, Dover Repreint, 1969.
[16] Fredricks, K.O., Ein Verfahren der Varitionsrechnung Nachv. Ges. Wiss. Göttingen, (1929), 13.
[17] Bellman, R. and W. Karush, 'On a New Functional Transform Analysis. The Maximum Transform'. Bull. Am. Math. Soc. 67 (1961), 501.
[18] Yourgrau, W. and S. Mandelstam, Variational Principles in Dynamics and Quantum Theory, W.B. Saunders Company, N.Y., 1968.
[19] Feynman, R.P., Rev. Mod. Phys. 20 (1948), 367.
[20] Schwinger, J., Phys. Rev. 82 (1951), 1916.
[21] Angel, E. and R. Bellman, Dynamic Programming and Partial Differential Equations, Academic Press, Inc., New York, 1972.
[22] Bellman, R., Introduction to Methods of Nonlinear Analysis, Academic Press, Inc., New York, Vols. I and II, 1970.
[23] Bellman, R. and R. Kalaba, Dynamic Programming and Modern Control Theory, Academic Press, Inc., New York, 1965.
[24] Bellman, R., Dynamic Programming, Princeton University Press, New Jersey, 1957.

[25] Bellman, R. and S. Dreyfus, Applied Dynamic Programming, Princeton University Press, N.J., 1961.
[26] Bellman, R., 'On a Class of Variational Problems', Quart. of Applied Math. 14 (1957), 353.
[27] Courant, R. and D. Hilbert, Methods of Mathematical Physics, Vol. I, Wiley & Sons, Inc., Interscience, New York, 1953.
[28] (a) Kohn, W., Phy. Rev. 74 (1948), 1763.
(b) Kato, T., Phy. Rev. 80 (1950), 475.

CHAPTER XI

DYNAMIC PROGRAMMING AND SOLUTION OF WAVE EQUATIONS

1. INTRODUCTION

In this chapter we use the dynamic programming techniques as applied to the various problems and arrive at the structure of the solutions of the second order equations without solving them. The variation diminishing properties of the Green's functin, the unimodal nature of the solutions of the Sturm-Liouville equations are derived in Sections 1 and 2. In Section 3, variational equations for the characteristic functions and characteristic values are obtained, treating one of the limits of the interval of integration as the imbedding parameter.

In Section 4, we deal with a method of obtaining bounds especially for the smaller eigenvalues of the Sturm-Liouville equation. Section 5 illustrates how one can convert a two point boundary value problem described by a second order differential equation into an initial value problem by obtaining a Cauchy system of equations for the Green's function. Employing dynamic programming methods, Section 6 arrives at the initial value equations for the Fredholm kernel of a Fredholm integral equation of the second kind.

The last few sections deal with scalar and matrix type of Riccati equations which play very important roles in many fields of engineering and mathematical physics.

2. PROPERTIES OF THE GREEN'S FUNCTION

Let us now consider interesting properties we are able to arrive at regarding the structure of the solutions of the second order linear equations without actually obtaining the soluitons. This can be done using the variational nature of the quadratic functional J(u) dealt with earlier. Let us now state without proof the variation diminishing properties of the Green's function associated with a suitable second order equation with a forcing term and zero end condition.

Let us begin with the one-dimensional case. We wish to observe that the solution of

$$u'' + g(t)u = h(t); \qquad u(0) = u(T) = 0, \tag{1}$$

is non-negative in [0, T] whenever h(t) is negative, provided that the quadratic functional

$$J(u) = \int_0^T (u'^2 - g(t)u^2)\,dt \tag{2}$$

is positive definite for functions satisfying the boundary conditions above.

It follows from this hypothesis that (1) is the Euler equation associated with the minimization of

$$J_1(u) = \int_0^T [u'^2 - g(t)u^2 + 2h(t)u]\,dt. \tag{3}$$

The solution of (1) furnishes the absolute minimum of Equation (3) over the class of functions such that $u' \in L^2(0, T)$, and the boundary conditions are satisfied. The reader is referred to the paper cited in reference [1].

Let us suppose that the stated result is not true which means that u, a continuous function is negative in some interval [a, b] (see Figure 1). Here a may be zero and b may be T.

Consider the new function v defined by

$$\begin{aligned} v &= u \quad t \notin [a, b], \\ v &= -u, \quad t \in [a, b]. \end{aligned} \tag{4}$$

The new function possibly has a discontinuous derivative at t = a and t = b, but nevertheless is such that $v' \in L^2(0, T)$ and v satisfies the boundary conditions.

Write

$$J_1(u) = \int_0^T = \int_a^b + \int_S , \tag{5}$$

where S = [0, T] - [a, b]. The integral over S is clearly unchanged by replacing u by v. On the other hand, since $h(t) \leq 0$ we see that the integral over [a, b] is decreased. Thus

$$J_1(v) < J_1(u) \ . \tag{6}$$

This is a contradiction to the fact that u provides the <u>absolute</u> minimum of J_1. Hence, $u \geq 0$ in [0, T].

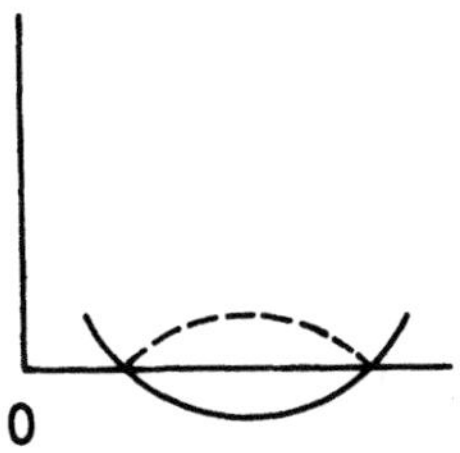

Fig. 1.

Since

$$u(t) = \int_0^T k(t, t_1)h(t_1)\, dt_1 , \tag{7}$$

and h(t) is an arbitrary negative function, this last fact implies that $k(t, t_1)$, the Green's function, is nonpositive for $0 \leq t, t_1 \leq T$, the desired result. Similarly in the same fashion we can show

$$u(t) = \int_0^T k(t, t_1)h(t_1)\, dt_1$$

cannot have more variations in sign than h(t) since the Green's function cannot have a greater number of oscillations than h(t). See reference [1]. This is the so-called variation diminishing property of the Green's function. This can be checked by actually computing the Green's function in several cases. Bellman's paper [1] gives more details treating Green's function as a smoothing function.

3. THE STURM OSCILLATION THEOREM AND UNIMODAL PROPERTIES

Let us now consider the problem of minimizing the functional

$$k(u) = \frac{\int u'^2\, dx}{\int p(x)u^2\, dx} \tag{8}$$

over all functions $u(x)$ such that $u(0) = u(a) = 0$. This leads to the characteristic value equation

$$u'' + \lambda p u = 0; \qquad u(0) = u(a) = 0 . \tag{9}$$

The minimum of $k(u)$ yields the smallest characteristic value. Now we take the equation

$$u'' + p(x)u = 0 \qquad u(0) = u(a) = 0 \tag{10}$$

with $p(x) > 0$, which has a solution which is positive for $0 < x < a$. We know that $\lambda = 1$ is the smallest characteristic value and that $u(x)$ is up to a normalization factor, the corresponding characteristic function.

Suppose we are given two equations

$$u'' + p(x)u = 0 \qquad u(0) = 0 \tag{11}$$

$$v'' + q(x)v = 0 \qquad v(0) = 0 \tag{12}$$

with $q(x) \geq p(x) > 0$. We wish to demonstrate that the first zero of $v(x)$ cannot occur after the first zero of $u(x)$.

Assume the situation were the reverse as indicated in Figure 2. Then from the property that unity is the lowest characteristic value,

$$1 = \min_{v} \frac{\int_0^b v'^2 \, dx}{\int_0^b q(x)v^2 \, dx} < \frac{\int_0^b w'^2 \, dx}{\int_0^b q(x)w^2 \, dx} , \tag{13}$$

where b is the first zero of v. w is another function defined over $[0, b]$.

Choose $w(x)$ in the following way

$$\left.\begin{aligned} w(x) &= u(x) \\ &= 0 \end{aligned}\right\} \quad \begin{aligned} &0 \leq x \leq a \\ &a \leq x \leq b \end{aligned} \tag{14}$$

Then in view of Equation (13) and (11) and the condition on q and p, we write

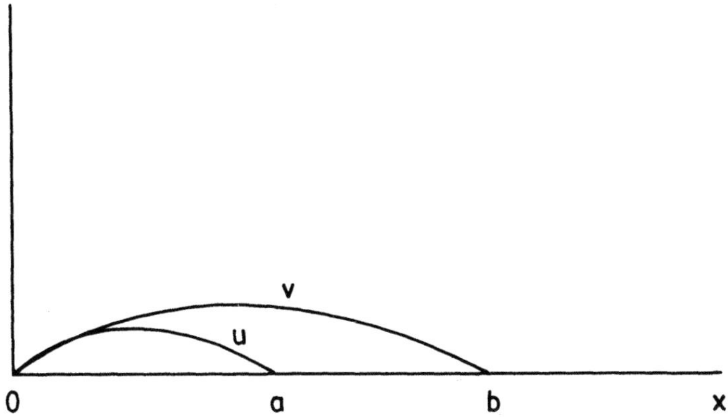

Fig. 2

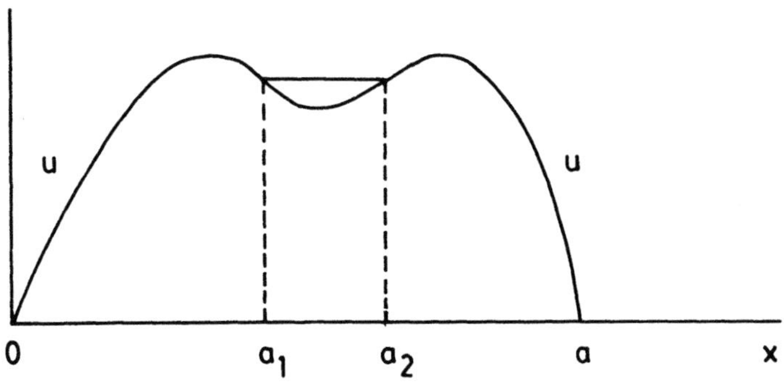

Fig. 3

$$1 < \frac{\int_0^a u'^2\, dx}{\int_0^a q(x)u^2(x)\, dx} \le \frac{\int_0^a u'^2\, dx}{\int_0^a p(x)u^2\, dx} = 1 \ . \tag{15}$$

This is a contradiction. Hence, the first zero of v cannot occur after the first zero of u (proven without obtaining the solutions u and v at all).

Similarly, we can demonstrate the unimodal property of the first-characteristic function u(x) which we take to be the solution of

$$u'' + p(x)u = 0 \qquad u(0) = u(a) = 0 \tag{16}$$

with the property $u(x) > 0;\ 0 < x < a$. Suppose it had the form as in Figure 3 with a double hump in the region $0 < x < a$.

The value of k(u) as defined in Equation (8) can be made smaller by using a different function v(x) which coincides with the value of u(x) at all points except in the region between $[a_2, a_1]$. In that region, let it take the value $u(a_1)$. As is obvious from Figure 3, v' in this region is zero. Hence,

$$\int_0^a v'^2\, dx < \int_0^a u'^2\, dx \tag{17}$$

$$\int_0^a p(x)v^2\, dx > \int_0^a p(x)u^2\, dx \ . \tag{18}$$

Hence,

$$k(v) < k(u) \ . \tag{19}$$

Since this is a contradiction to the minimizing properties of u, u(x) cannot have two humps as shown in Figure 3 [2].

4. CHARACTERISTIC VALUES AND CHARACTERISTIC FUNCTIONS

The functional equation technique of the dynamic programming method can be applied to derive variational relations for the kernel functions and Green's functions. Let us consider the general equation of the type

$$(p(x)u')' + [r(x) + \lambda q(x)]u = 0$$
$$u(a) = u(1) = 0 \; . \tag{20}$$

We want to obtain the variational equations for the resolvent operator as a function of 'a'. Utilizing the meromorphic nature of the operator as a function of λ, we are able to obtain the variational equation for characteristic values and functions as shown below.

We first consider the boundary value problem

$$(p(x)u')' + q(x)u = v(x) \qquad a < x < 1$$
$$u(a) = 0 \qquad u(1) + \alpha u'(1) = 0 \; . \tag{21}$$

We shall assume that p, q, and v are continuous functions on the closed interval [a, 1]. If the corresponding Sturm-Liouville problem with q replaced by λq and v by 0 does not have 1 as a characteristic value, then the unique solution of (21) can be represented in the form

$$u(x) = \int_a^1 K(x, y, a)v(y)\, dy \; . \tag{22}$$

The function K(x, y, a) is called the Green's function for the boundary value problem (21). We wish to study the dependence of K upon 'a' by means of the functional equation method of dynamic programming.

We imbed (21) in the system

$$(p(x)u')' + q(x)u = v(x), \qquad a < x < 1$$
$$u(a) = c, \qquad u(1) + \alpha u'(1) = 0 \; . \tag{23}$$

A solution of this system can be expressed in the form

$$u(x) = u_0(x) + c\phi(x) \; , \tag{24}$$

where $u_0(x)$ is a solution of (21) and hence equal to the right side of (22) while $\phi(x)$ is a solution of the system

$$(p(x)\phi')' + q(x)\phi = 0, \quad \phi(a) = 1, \quad \phi(1) + \alpha\phi'(1) = 0 . \tag{25}$$

We also consider a variational problem associated with the system (23), the problem of maximizing J(u, v) over all u for which u(a) = c, where

$$j(u, v) = \int_a^1 (q(x)u^2 - p(x)u'^2 - 2uv(x))\, dx - \frac{p(1)}{\alpha}[u(1)]^2 . \tag{26}$$

This variational problem has the property that any u yielding the maximum is a solution of the system (23). To prove this, we replace u by $u + \varepsilon\eta$ in (26) where $\eta(x)$ is a function such that $\eta(a) = 0$. We obtain

$$J(u+\varepsilon\eta, v) = J(u, v) + 2\varepsilon\left\{\int_a^1 (qu\eta - pu'\eta' - v\eta)\, dx - \right.$$

$$\left. - \frac{p(1)}{\alpha} u(1)\eta(1)\right\} + \varepsilon^2\left[\int_a^1 (q\eta^2 - p\eta'^2)\, dx - \right.$$

$$\left. - \frac{p(1)}{\alpha}\eta(1)^2\right] . \tag{26a}$$

In order to have u yield the maximum, the ε term must vanish for all functions η for which $\eta(a) = 0$. Hence

$$\int_a^1 \left\{[qu + (pu')' - v(x)]\eta\right\} dx - \frac{p(1)}{\alpha}\eta(1)u(1) + \alpha u'(1)) = 0. \tag{26b}$$

Consequently, any u yielding the maximum must satisfy the boundary condition at x = 1 given by Equation (23) and must satisfy the differential equation in the interior of the interval [a, 1]. Hence, a solution of the variational problem, provides a solution of the boundary value problem (23).

In order to use the variational approach one must make an assumption on p and q sufficient to guarantee the existence of a maximum. It is sufficient, for example, to assume that $p(x)$ is positive on the closed interval $[a, 1]$ and that the smallest characteristic value of the Sturm-Liouville problem

$$(pu')' + \lambda qu = 0 \quad u(a) = 0 \quad u(1) + \alpha u'(1) = 0$$

is larger than 1. If $q(x)$ is uniformly positive, $d \geq q(x) \geq 0$, over $0 \leq x \leq a$, this condition holds if a is sufficiently small, or if d is sufficiently large. This assumption guarantees the existence of a unique maximum of $J(u, v)$. By analytic continuation the results can be freed from this restrictive hypothesis.

A. Dynamic Programming Approach

We now let

$$f(a, c) = \max_{u(a) = c} J(u, v) \tag{27}$$

and derive a partial differential equation for f by means of a technique of dynamic programming.

We regard u as describing a policy. The variable c describes the state of the system at a. The result of following the policy u for a time interval $[a, a+\Delta]$ is to transform c into a new initial state $u(a+\Delta)$ for the interval $[a+\Delta, 1]$. According to the principle of optimality, an optimal policy u over the interval $[a, 1]$ with initial state c must have the property that it is an optimal policy over the subinterval $[a+\Delta, 1]$ starting with initial state $u(a+\Delta)$. Translated into a formula, the principle of optimality yields the equation

$$f(a, c) = \max_{u(a)=c} \{f[(a+\Delta), u(a+\Delta)] + \int_a^{a+\Delta} [q(x)u^2(x) - p(x)u'^2(x) - 2u(x)v(x)]\, dx\}. \tag{27a}$$

We proceed formally, assuming that $f(a, c)$ has continuous partial derivatives and that the maximizing u has a continuous derivative. Then as $\Delta \to 0$

$$f(a+\Delta,\ u(a+\Delta)) = f(a,\ c) + \frac{\partial f}{\partial a}(a,\ c)\Delta + \frac{\partial f}{\partial c}(a,\ c)u'(a)\Delta + o(\Delta)\ .$$

Consequently, as $\Delta \to 0$, we have

$$-\frac{\partial f}{\partial a}(a,\ c) = \max_{u(a)=c}\left\{\frac{\partial f}{\partial c}(a,\ c)u'(a) + q(a)c^2 - p(a)u'^2(a) - \right.$$

$$\left. - 2cv(a) + o(1)\right\}.$$

The quadratic $(\partial f/\partial c)u' - p(a)u'^2$ takes its maximum value

$$u' = \frac{1}{2p(a)}\frac{\partial f}{\partial c}(a,\ c).$$

Hence, we obtain the partial differential equation

$$-\frac{\partial f}{\partial a}(a,\ c) = \frac{1}{4p(a)}\left[\frac{\partial f}{\partial c}(a,\ c)\right]^2 - 2cv(a) + c^2q(a)\ . \qquad (27b)$$

B. Variation of the Green's Function

Let u be the function which maximizes $J(u,\ v)$ for given a and c. By using (20) and (25), we can find an equation which connects $f(a,\ c)$ with the Green's function $K(x,\ y,\ a)$. We have

$$f(a,\ c) = \int_a^1 (qu^2 - pu'^2 - 2uv)\,dx - \frac{p(1)}{\alpha}(u(1))^2$$

$$= \int_a^1 u(qu + (pu')' - v)\,dx - \int_a^1 uv\,dx - u(1)p(1)u'(1) +$$

$$+ u(a)p(a)u'(a) - \frac{p(1)}{\alpha}[u(1)]^2$$

$$= -\int_a^1 uv\,dx + cp(a)u'(a) - \frac{p(1)}{\alpha}u(1)[u(1)+\alpha u'(1)]$$

$$= -\int_a^1 u_0v\,dx - c\int_a^1 \phi v\,dx + cp(a)u_0'(a) + c^2p(a)\phi'(a). \qquad (28)$$

Here we have used the equation for the solution u, given by (24).
Using the fact that u_0 satisfies (21) and ϕ satisfies (25), we obtain by integration by parts

$$\int_a^1 \phi v \, dx = \int_a^1 \phi[(pu_0')' + qu_0] \, dx$$

$$= \int_a^1 u_0[(p\phi')' + q\phi] \, dx + \phi(1)p(1)u_0'(1)$$
$$- \phi(a)p(a)u_0'(a)$$
$$- \phi'(1)p(1)u_0(1)$$
$$+ \phi'(a)p(a)u_0(a) \, . \quad (28a)$$

Since $u_0(a) = 0$, $\phi(a) = 1$, and u and ϕ satisfy the same homogeneous boundary condition at 1, we conclude that

$$\int_a^1 \phi v \, dx = -p(a)u_0'(a) + [\phi(1)p(1)u_0'(1) - \phi'(1)p(1)u_0(1)] \quad (29)$$

$$= -p(a)u_0'(a) \, .$$

Thus, the second and third term on the last line of (28) are equal. It turns out to be more convenient to use the integral expression. We have

$$f(a, c) = -\int_a^1 u_0 v \, dx - 2c\int_a^1 \phi v \, dx + c^2 p(a)\phi'(a)$$

$$= -\int_a^1\int_a^1 K(x, y)v(x)v(y) \, dx \, dy - \quad (30)$$

$$- 2c\int_a^1 \phi v \, dx + c^2 p(a)\phi'(a) \, .$$

We observe that we can express ϕ in terms of a partial derivative of K. By (29), we have

$$\int_a^1 \phi(y)v(y)\,dy = -p(a)u_0'(a) = -p(a)\int_a^1 \frac{\partial K}{\partial x}(a, y, a)v(y)\,dy$$

for all continuous functions v. Hence,

$$\phi(y) = -p(a)\frac{\partial K}{\partial x}(a, y, a) = -p(a)\left.\frac{\partial K(x, y, a)}{\partial x}\right|_{x=a}. \quad (31)$$

Combining the expression for f(a, c) given by (30) with the partial differential Equation (27b), we obtain upon equating terms independent of c,

$$\int_a^1\int_a^1 \frac{\partial K}{\partial a}(x, y, a)v(x)v(y)\,dx\,dy = \frac{4}{4p(a)}\left[\int_a^1 \phi(x)v(x)\,dx\right]^2$$

$$= \frac{1}{p(a)}\int_a^1\int_a^1 \phi(x)\phi(y)v(x)v(y)\,dx\,dy\ .$$

Now if we 'equate coefficients' of v(x)v(y) we obtain the relation

$$\frac{\partial K}{\partial a}(x, y, a) = \frac{1}{p(a)}\phi(x)\phi(y) = p(a)\frac{\partial K}{\partial y}(x, a, a)\frac{\partial K}{\partial x}(a, y, a)\ . \quad (32)$$

This formal equating of coefficients can be justified with the use of the symmetry of the Green's function and an assumption such as continuity of

$$\frac{\partial K}{\partial a}(x, y, a) - \frac{1}{p(a)}\phi(x)\phi(y)\ .$$

For proof, see Reference [3].

In consideration of the boundary value problem (23), we have not considered the boundary conditions u(1) = 0 or u'(1) = 0. The condition u(1) = 0 corresponds formally to $\alpha = 0$; the condition u'(1) = 0 corresponds formally to $\alpha = \infty$

in the condition $u(1) + \alpha u'(1) = 0$. If in the variational problem of maximizing $J(u, v)$ we omit the term

$$\frac{-p(1)[u(1)]^2}{\alpha}$$

and add the constraint $u(1) = 0$ or $u'(1) = 0$, respectively, the development proceeds as before, except that all terms which involve α vanish. The final results obtained are the same.

C. Analytic Continuation

In order to extend the foregoing approach to the general case, we shall employ the method of analytic continuation.

Consider the equation

$$(pu')' + (z+q(x))u = v;\ u(a) = c,\ u(1) + \alpha u'(1) = 0, \tag{33}$$

where z is a positive quantity chosen large enough so that $-z + q(x) \leq 0$ in $[a, 1]$.

Furthermore, we know from the Sturmian comparison theorems that z can be chosen large enough so that $\lambda = 1$ is not a characteristic value of the Sturm-Liouville problem

$$(pu')' + \lambda(z+q(x))u = v,\ u(a) = 0,\ u(1) + \alpha u'(1) = 0\ . \tag{34}$$

It follows that the inhomogeneous equation

$$(pu')' + (z + q(x))u = v,\ u(a) = 0,\ u(1) + \alpha u'(1) = 0 \tag{35}$$

will then have a unique solution.

The Green's function associated with this problem will be a function of z. If we can show that this function is a meromorphic function of z, it will follow that the relations originally obtained under the assumption that z is sufficiently large will actually be valid for all z distinct from a set of poles.

In particular, the relations will be valid for $z = 0$, provided that $\lambda = 1$ is not a characteristic value of the equation

$$(pu')' + \lambda q(x)u = 0,\ u(a) = 0,\ u(1) + \alpha u'(1) = 0\ . \tag{36}$$

Generally speaking, the relations will be valid whenever they make sense.

D. Analytic Character of Green's Function

Although it follows from well-known results in the theory of linear differential equations that the Green's function is a meromorphic function of z, we shall outline a proof here for the sake of completeness.

To simplify the notation let us write $\sigma(u) = 0$ for the boundary condition $u(1) + \alpha u'(1) = 0$. Furthermore, we observe that the following considerations also cover the boundary condition $u'(1) = 0$, so that we could also take $\sigma(u) = u'(1)$.

To begin with, the two solutions of

$$(pu')' + (z + q(x))u = 0 \tag{37}$$

determined by the initial conditions

$$u_1(a) = 1,\ u_1'(a) = 0,\ u_2(a) = 0,\ u_2'(a) = 1 \tag{38}$$

are entire functions of z for x in [0, 1]. Also for fixed x, their derivatives u_1' and u_2' are entire functions of z. Hence, in particular, $\sigma(u_2)$ is an entire function of z which does not vanish identically.

Concerning p(x) and q(x), we are assuming, as before, that $p(x) \geq d_1 > 0$ in [0, 1], that it has an integrable derivative, and that q(x) is integrable in [a, 1], and uniformly bounded.

The general solution of

$$(pu')' + (z + q(x))u = v \tag{39}$$

can be written

$$u = c_1u_1(x) + c_2u_2(x) - \frac{1}{p(a)}\int_a^x [u_1(x)u_2(y) - u_1(y)u_2(x)]v(y)\,dy. \tag{40}$$

Imposing the boundary conditions

$$u(a) = 0;\quad \sigma(u) = 0$$

we see that this particular solution has the form

$$u = \int_a^1 K(x, y)v(y)\, dy ,$$

where K(x, y), the desired Green's function, is given by

$$K(x, y) = \frac{1}{p(a)} \left[\frac{\sigma(u_1)u_2(y) - u_1(y)\sigma(u_2)}{\sigma(u_2)}\right] u_2(x); \quad a < y < x$$

$$= \frac{1}{p(a)} \left[\frac{\sigma(u_1)u_2(x) - u_1(x)\sigma(u_2)}{\sigma(u_2)}\right] u_2(y); \quad a < x < y .$$

We see that for any x and y in [a, 1], K(x, y) and its partial derivatives with respect to x, y and a are meromorphic functions of z. Also, we observe that this formula provides a proof for the symmetry of the Green's function, a property which we have used earlier.

E. Alternate derivation of Expression for $\sigma(x)$

The explicit representation for K(x, y) given above shows that

$$\left.\frac{\partial K(x, y)}{\partial x}\right|_{x=a} = \frac{1}{p(a)} \left[\frac{\sigma(u_1)u_2(y) - u_1(y)\sigma(u_2)}{\sigma(u_2)}\right]$$

It follows then that $-p(a)\ \partial K/\partial x(a, y, a)$ is the solution of the two-point boundary value problem

$$(pu')' + qu = 0 \quad u(a) = 1 \quad \sigma(u) = 0.$$

Hence, we have

$$\phi(y) = -p(a)\, \frac{\partial K}{\partial x}\, (a, y, a),$$

as in Equation (31). The reader can look up references [4] and [5] for extensions of these results.

F. Variation of Characteristic Values and Characteristic Functions

Consider the Sturm-Liouville problem

$$(pu')' + (q(x) + \lambda r(x))u = v(x); \quad a < x < 1 \tag{41}$$

$$u(a) = 0; \qquad u(1) + \alpha u'(1) = 0.$$

This problem has a unique solution, when λ is not a characteristic value, which is given by

$$u(x) = \int_a^1 R(x, y, \lambda, a)v(y)\,dy .$$

The function R, called the resolvent is a meromorphic function of λ with poles at the characteristic values $\lambda_1, \lambda_2, \lambda_3, \cdots$. If $\Psi_k(x)$ is the characteristic function associated with the characteristic value λ_k, then R has the representation

$$R(x, y, \lambda, a) = \sum_k \frac{\Psi_k(x)\Psi_k(y)}{\lambda - \lambda_k} . \tag{42}$$

The problem (41) is obtained from (21) by replacing q(x) by $q(x) + \lambda r(x)$. From (32) we obtain the equation

$$\frac{\partial R(x, y, \lambda, a)}{\partial a} = p(a) \frac{\partial R}{\partial x}(a, y, \lambda, a) \frac{\partial R}{\partial y}(x, a, \lambda, a) . \tag{43}$$

This equation is valid for all $\lambda \neq \lambda_k$ without any assumptions on q(x) and r(x) except continuity as can be shown by analytic continuation.

Combining (42) with (43), we find

$$\sum_{k=1}^{\infty} \left[\frac{\Psi_k(x)\Psi_k(y)}{(\lambda-\lambda_k)^2} \frac{\partial \lambda_k}{\partial a} + \frac{\Psi_k(x) \dfrac{\partial \Psi_k(y)}{\partial a} + \Psi_k(y) \dfrac{\partial \Psi_k(x)}{\partial a}}{\lambda - \lambda_k} \right] =$$

$$= p(a) \left(\sum_{k=1}^{\infty} \frac{\Psi_k'(a)\Psi_k(y)}{(\lambda-\lambda_k)} \right) \left(\sum_{k=1}^{\infty} \frac{\Psi_k(x)\Psi_k'(a)}{\lambda-\lambda_k} \right) .$$

Letting $\lambda \to \lambda_k$ and equating coefficients of $(\lambda - \lambda_k)^{-2}$ we find

$$\frac{\partial \lambda_k}{\partial a} = p(a)(\Psi_k'(a))^2 . \tag{44}$$

By equating coefficients of $(\lambda - \lambda_k)^{-1}$ we find

$$\frac{\partial \Psi_k(x)}{\partial a} = p(a)\Psi_k'(a)\left(\sum_{j \neq k} \frac{\Psi_j(x)\Psi_j'(a)}{\lambda - \lambda_j} \right) . \tag{45}$$

We can also derive expressions for the variation of the sums

$$S_n(x, y, a) = \sum_{k=1}^{\infty} \frac{\Psi_k(x)\Psi_k(y)}{\lambda_k^n} \qquad (n = 1, 2, \cdots) . \tag{46}$$

Since

$$\frac{1}{\lambda - \lambda_k} = - \frac{1}{\lambda_k} - \frac{\lambda}{\lambda_k^2} - \frac{\lambda^2}{\lambda_k^3} - \cdots \tag{47}$$

we have

$$R(x, y, \lambda, a) = - \sum_{n=1}^{\infty} S_n(x, y, a)\lambda^{n-1} . \tag{48}$$

Consequently, from Equation (43) we see that

$$\frac{\partial S_n(x, y, a)}{\partial a} = -p(a) \sum_{\substack{j+k=n+1 \\ 1 \leq j \leq n \\ 1 \leq k \leq n}} \sum \frac{\partial S_j(a, y, a)}{\partial x} \frac{\partial S_k(x, a, a)}{\partial y} . \tag{49}$$

This can be extended to the matrix case also. The vector matrix equation* to be considered is

---

* The vector and matrix nature of the entities in the equations can be easily understood from the context without any special indications.

$$\frac{d}{dt}\left[P(t)\frac{d}{dt}X(t)\right] + Q(t)X(t) = Y(t);\ x \le t \le 1$$
$$X(a) = c \qquad \dot{X}(1) + BX(1) = 0 \tag{50}$$

where P, Q, and B are n × n matrices, $\vec{X}$, $\vec{Y}$ and $\vec{c}$ are n dimensional vectors. We assume P, Q, and Y are continuous on the closed interval [a, 1], and P(t) has an inverse for each t. Also, we assume that P, Q, and P(1)B are symmetric matrices so that the system is self-adjoint.

Associated with this boundary value problem, there is the variational problem of minimizing J(X, Y)

$$J(X, Y) = \int_a^1 \{(QX, X) - (P\dot{X}, \dot{X}) - 2(Y, X)\}\, dt - (P(1)BX(1),\ X(1)) \tag{51}$$

subject to the condition $\vec{x}(a) = c$. We consider the functional

$$f(a, c) = \underset{X(a) = c}{\mathrm{Max}}\ J(X, Y)\ .$$

The procedure then follows in exactly the same fashion as in the scalar case with few modifications and we obtain for the Green's matrix K(t, s), which corresponds to the scalar Green's function and has the variational equation

$$\frac{\partial K}{\partial a}(t, s, a) = \Phi(t)P^{-1}(a)\Phi^*(s)\ , \tag{52}$$

where Φ is the solution of the homogeneous equation

$$\frac{d}{dt}[P(t)\ \frac{d}{dt}\ \Phi(t)] + Q(t)\Phi(t) = 0 \tag{53}$$

$$\Phi(a) = I \qquad \Phi(1) + B\dot{\Phi}(1) = 0\ . \tag{54}$$

For other results the reader is referred to the literature cited.

5. DETERMINATION OF CHARACTERISTIC VALUES OF THE STURM-LIOUVILLE EQUATION

We describe in this section some of the methods of obtaining the characteristic values of the Sturm-Liouville equations, especially the smaller eigenvalues, by finding upper and lower bounds for the minimum eigenvalue λ. We follow the methods devised in the paper by Bellman [6].

Let us start with the equation

$$u'' + \lambda a(t)u = 0; \quad u(0) = u(1) = 0 . \tag{55}$$

It will be clear from what follows that the methods we discuss can be applied to questions of this type involving quite general boundary conditions, as long as the interval is finite.

There are, at present, a number of powerful techniques available for treating problems of this genre, based upon variational techniques, and upon matrix techniques applied to a finite difference version of the foregoing differential equation.

The variational approach depends upon the fact that if a(t) satisfies a reasonable condition such as

$$0 < a^2 \leq a(t) \leq b^2 < \infty; \quad 0 \leq t \leq 1 \tag{56}$$

then the characteristic values, $\lambda_1 < \lambda_2 < \cdots$, are the respective relative minima of the functional

$$J(u) = \int_0^1 u'^2 \, dt \Big/ \int_0^1 a(t)u^2 \, dt \tag{57}$$

as u ranges over the space of functions for which the integrals exist and for which u(0) = u(1) = 0.

In particular,

$$\lambda_1 \leq \int_0^1 u'^2 \, dt \Big/ \int_0^1 a(t)u^2 \, dt \tag{58}$$

for all functions u(t) satisfying the prescribed boundary conditions. We thus have a means of obtaining upper bounds for λ_1 which turn out to be remarkable accurate even for simple choices of trial functions u(t).

Another method is based upon using equatios of the form

$$u_{n+2} - 2u_{n+1} + u_n + \lambda\Delta^2 a_n u_n = 0 \tag{59}$$

u(0) = u(N) = 0, and applying any of a number of methods used to derive the characteristic roots and vectors of a symmetric matrix. For a detailed discussion of these methods, and others, we refer to the book by Collatz [7].

There is, however, a significant difference between a problem of this type, and the Sturm-Liouville problem described above. This is due to the fact that it is quite easy to find asymptotic solutions to (55) for large λ, and thus, approximate expressions for the higher characteristic values.

Let, for simplicity of notation, $a(t) = q^2(t)$; then the Liouville transformation $s = \int_0^t q(t_1)\, dt_1$, converts

$$u'' + \lambda q^2(t) u = 0 \tag{60}$$

into

$$\frac{d^2u}{ds^2} + \frac{q'}{q}\frac{du}{ds} + \lambda u = 0 . \tag{61}$$

The further transformation

$$v = u\sqrt{q} = u \exp\left\{\frac{1}{2}\int \frac{q'}{q}\, ds\right\} \tag{62}$$

converts (61) into

$$\frac{d^2v}{ds^2} + \left[\lambda - \frac{1}{2}\frac{d}{ds}\left(\frac{q'}{q}\right) - \left(\frac{q'}{q}\right)^2\right]v = 0. \tag{63}$$

The new boundary conditions are

$$v(0) = 0; \quad v\left(\int_0^1 q(t)\, dt\right) = 0 . \tag{64}$$

Writing (63) in the form

$$v''(s) + (\lambda + b(s))v(s) = 0 . \tag{65}$$

We know that we can find asymptotic developments for $v(s)$ starting from the integral equation

$$v(s) = c_1 \cos \lambda^{1/2} s + c_2 \sin \lambda^{1/2} s - \int_0^s \frac{\sin \lambda^{1/2}(s-r)}{\lambda^{1/2}} b(r)v(r)\,dr \tag{66}$$

and iterating. Approximate values of λ are now determined by means of the constraint $v\left(\int_0^1 q(t)\,dt\right) = 0$. Thus, the higher characteristic values have the principal term

$$\lambda_n \simeq \frac{n^2\pi^2}{\left[\int_0^1 q(t)\,dt\right]^2} . \tag{67}$$

To obtain more precise results, we can use further terms of the asymptotic series derived from (66) and we can combine this with the numerical integration of (55).

It follows from these considerations that the greatest difficulty is experienced in obtaining accurate estimations of the first characteristic value. In many investigations this is all that is desired.

We wish to deal with a method in Reference [6] suitable for digital computation.

Consider the linear differential equation

$$u'' + \lambda a(t)u = 0, \quad u(0) = 0 . \tag{68}$$

Let us assume $u'(0) = 1$. The value of $u'(0)$ can be anything else. This does not alter the considerations here since the final solution gets multiplied by that value. We have the value $u(1)$ to be zero at the end $x = 1$. The solution of this initial value problem may be obtained over $0 \leq t \leq 1$ as a power series in λ in the form

$$u = t + \sum_{n=1}^{\infty} u_n(t)\lambda^n , \tag{69}$$

where the sequence of coefficient functions $\{u_n(t)\}$, $n = 1, 2, \ldots$, may be determined by means of the recurrence relations

$$u_0(t) = t,$$
$$u_n(t) = -\int_0^t (t-s)u_{n-1}(s)a(s)\,ds, \quad n = 1, 2, \ldots, . \tag{70}$$

It is easy to see that u, as defined by (69), is an analytic function of λ for all finite λ for $0 \leq t \leq 1$. The roots of the equation

$$f(\lambda) = u(1) = 1 + \sum_{n=1}^{\infty} u_n(1)\lambda^n = 0 \tag{71}$$

are the desired characteristic values.

Referring to the equation in (68), we easily see from the form of the solution in Equation (69) that

$$|u(t)| \leq e^{k|\lambda|^{1/2}} \tag{72}$$

for $0 \leq t \leq 1$, where k is a constant. Consequently, the Weierstrass factorization of $f(\lambda)$ takes the form

$$f(\lambda) = \prod_{i=1}^{\infty} (1 - \lambda/\lambda_i) . \tag{73}$$

As we know, $\lambda_n = 0(n^2)$ as $n \to \infty$, in view of the assumptions we have made concerning a(t) in (57).

Our aim is now, by following the technique used by Newton to relate the sums of the powers of the roots and the elementary symmetric functions which are the coefficients, to obtain relations for the sums

$$b_r = \sum_{i=1}^{\infty} 1/\lambda_i^r \quad r = 1, 2, \ldots , \tag{74}$$

in terms of the coefficients $u_n(1)$.

It is clear that

$$\log f(\lambda) = \sum_{i=1}^{\infty} \log\left(1 - \frac{\lambda}{\lambda_i}\right) = -\sum_{r=1}^{\infty} \frac{\lambda^r}{r} \sum_{i=1}^{\infty} \frac{1}{\lambda_i^r} = -\sum_{r=1}^{\infty} \lambda^r \frac{b_r}{r} \tag{75}$$

for $|\lambda| < \lambda_1$.

It is important then to obtain the coefficients of the expression of log f(λ). Although this can be done directly, it is easier to proceed as follows. Write

$$\log f(\lambda) = \sum_{k=1}^{\infty} c_k \lambda^k . \tag{76}$$

Then

$$\frac{f'(\lambda)}{f(\lambda)} = \sum_{k=1}^{\infty} k c_k \lambda^{k-1} , \tag{77}$$

whence

$$\sum_{n=1}^{\infty} n u_n(1)\lambda^{n-1} = \left(\sum_{k=1}^{\infty} k c_k \lambda^{k-1}\right)\left(1 + \sum_{n=1}^{\infty} u_n(1)\lambda^n\right) , \tag{78}$$

whence we obtain the well-known recurrence relations

$$n u_n = n c_n + \sum_{k=1}^{n-1} k c_k u_{n-k} \ ; \ u_0(t) = t \tag{79}$$

u_n can be simultaneously determined in a computer. These permit us to calculate the c_n in a very simple fashion once the sequence $\{u_n(1)\}$ has been determined, and thence the b_n.

Let us now show that the sequence $\{b_k\}$ can be used to obtain sequences which converge monotonically from above and below to the first characteristic value λ_1. We then can prove that

$$\frac{b_k}{b_{k+1}} > \lambda_1 > \frac{1}{b_k^{1/k}} . \tag{80}$$

The sequence $\{b_k/b_{k+1}\}$ is monotone decreasing; the sequence $\{1/b_k^{1/k}\}$ is monotone increasing, and

$$\lambda_1 = \lim_{k\to\infty} \frac{b_k}{b_{k+1}} = \lim_{k\to\infty} \frac{1}{b_k^{1/k}} . \tag{81}$$

The monotone character of the ration b_k/b_{k+1} follows directly from Schwarz's inequality, since from Equation (74)

$$b_k^2 = \left[\sum_{i=1}^{\infty} \frac{1}{\lambda_i^k}\right]^2 = \left(\sum_{i=1}^{\infty} \frac{1}{\lambda_i^{\frac{k+1}{2}} \lambda_i^{\frac{k-1}{2}}}\right)^2$$
$$\leq \left(\sum_{i=1}^{\infty} \frac{1}{\lambda_i^{k+1}}\right)\left(\sum_{i=1}^{\infty} \frac{1}{\lambda_i^{k-1}}\right) = b_{k+1} b_{k-1} \ . \tag{82}$$

Therefore $b_k/b_{k+1} \leq b_{k-1}/b_k$. The monotone behavior of $b_k^{1/k}$ is a consequence of the well-known inequality

$$\left(\sum_{i=1}^{\infty} x_i\right) \geq \left(\sum_{i=1}^{\infty} x_i^2\right)^{1/2} \geq \left(\sum_{i=1}^{\infty} x_i^3\right)^{1/3} \geq \cdots \tag{83}$$

for any set of nonnegative x_i.

Since

$$b_k^{1/k} = \left(\sum_{i=1}^{\infty} \frac{1}{\lambda_i^k}\right)^{1/k}$$

we have the monotone increasing property for $1/b_k^{1/k}$. The proof of the limiting relation is clear since

$$\frac{b_k}{b_{k+1}} = \frac{\frac{1}{\lambda_1^k}\left[1 + \left(\frac{\lambda_1}{\lambda_2}\right)^k + \cdots\right]}{\frac{1}{\lambda_1^{k+1}}\left[1 + \left(\frac{\lambda_1}{\lambda_2}\right)^{k+1} + \cdots\right]}$$
$$= \lambda_1\left[1 + \left(\frac{\lambda_1}{\lambda_2}\right)^k - \left(\frac{\lambda_1}{\lambda_2}\right)^{k+1} + \cdots\right] . \tag{84}$$

We see that

$$\frac{b_k}{b_{k+1}} - \lambda_1 \simeq \lambda_1 \left(\frac{\lambda_1}{\lambda_2}\right)^k \tag{85}$$

for large k.

Similarly,

$$b_k^{1/k} = \frac{1}{\lambda_1}\left[1 + \left(\frac{\lambda_1}{\lambda_2}\right)^k + \cdots\right]^{1/k} \simeq \frac{1}{\lambda_1}\left[1 + \frac{1}{k}\left(\frac{\lambda_1}{\lambda_2}\right)^k + \ldots\right] \tag{86}$$

for large k. Hence we can realize the meaning of Equations (80) and (81).

It is to be expected that b_k/b_{k+1} will furnish a better approximation to λ_1 for large k.

For the case where $a(t) \equiv 1$, $\lambda_1/\lambda_2 = \frac{1}{4}$, the rate of convergence of these sequences will not be too rapid. In those situations there are two things we can do to obtain more accurate estimations of λ_1. In the first place, we can use the root-squaring technique. Since

$$f(\lambda) = \prod_{i=1}^{\infty} (1 - \lambda/\lambda_i) \tag{87}$$

We see that

$$f_1(\lambda) = f(\lambda^{1/2})f(-\lambda^{1/2}) = \prod_{i=1}^{\infty}\left(1 - \frac{\lambda}{\lambda_i^2}\right). \tag{88}$$

Using the power series development for $f_1(\lambda)$ we obtain a sequence $\{b_k'\}$ with

$$\lim_{k\to\infty} \frac{b_k'}{b_{k+1}'} = \lambda_1^2 \tag{89}$$

and a rate of convergence depending upon $(\lambda_1/\lambda_2)^2$.

Alternatively, once we have an estimate for λ_1 with an accuracy of 1 in 10^{-s}, we can then turn to the power series for $f(\lambda)$ and use the Newton-Raphson approximation techniques,

$$\lambda_1^{(n+1)} = \lambda_1^{(n)} - f(\lambda_1^{(n)})/f'(\lambda_1^{(n)}). \tag{90}^*$$

* $\lambda^{(n)}$ is the nth approximation for λ.

This will yield a further approximation with accuracy of essentially 1 in 10^{-2s}. Continued use of this technique is limited only by the number of $u_n(1)$ which are computed, and the accuracy of this computation. There is no difficulty involved in using this technique since we know from theoretical considerations that the roots of $f(\lambda)$ are simple.

Similar upper bounds can be otbained for the products $\Pi_{i=1}^{R+1} \lambda_i$; $R = 1, 2, \cdots$.

Consider the determinant

$$b_k^{(R)} = \begin{vmatrix} b_k, & b_{k+1} & \cdots & b_{k+R} \\ b_{k+R} & \cdots\cdots & & b_{k+2R} \end{vmatrix}; \quad R = 1, 2, \cdots . \tag{91}$$

It is not difficult to show that

$$\lim_{k\to\infty} \frac{b_k^{(R)}}{b_{k+1}^{(R)}} = \lim_{k\to\infty} (b_k^{(R)})^{-1/k} = \lambda_1\lambda_2 \cdots \lambda_{R+1}. \tag{92}$$

To show that

$$\frac{b_k^{(R)}}{b_{k+1}^{(R)}} > \frac{b_{k+1}^{(R)}}{b_{k+2}^{(R)}}, \quad k = 1, 2, \cdots \tag{93}$$

for $R = 1, 2, \cdots$, we use the well-known fact that the matrix

$$B_k^{(R)} = \begin{pmatrix} b_k & b_{k+1} \cdots & b_{k+R} \\ b_{k+R} & \cdots\cdots & b_{k+2R} \end{pmatrix} \tag{94}$$

is positive definite for all k and R, and hence that $(B_k^{(R)})^{-1}$ is positive definite.

The sequence $(B_k^{(R)})^{-1/k}$ does not seem to have any simple monotonicity properties.

Since we can obtain limits for the products of λ_k's as above, we can obtain sharp estimates for higher eigenvalues, $\lambda_2\lambda_3 \cdots$ etc, successively. This method has been used to obtain eigenvalues of equations of the type

$$u'' + \lambda(1+t)u = 0 \qquad u(0) = u(1) = 0 \,.$$

The numerical results are in excellent agreement with values obtained by analytical means. The reader's attention is drawn to references [6] and [8].

6. ANOTHER TYPE OF CAUCHY SYSTEM FOR THE GREEN'S FUNCTION AND THE SOLUTION OF THE TWO POINT BOUNDARY VALUE PROBLEM

In an earlier section, equations for the Green's function were obtained by imbedding in the length of the interval of the region in which the solution of the differential equation is to be found. We can arrive at a system of initial value equations if the imbedding is done with regard to another parameter occurring in the differential equations. This Cauchy system for the Green's function leads to the solution of the two point boundary value problem. Here we follow the method developed in the papers by Huss et al. [9].

Consider the linear two point boundary value problem

$$\begin{aligned} &\ddot{x} + \lambda q(t)x(t) = -g(t); \quad 0 \le t \le 1 \\ &x(0) = 0; \; x(1) = c \end{aligned} \tag{95}$$

assuming that a unique solution exists for λ, $0 \le \lambda \le \lambda_1$. We introduce two new functions u and w that are the solutions of linear two point boundary value problems

$$\begin{aligned} &\ddot{u} + \lambda q u = 0 \quad 0 \le t \le 1 \\ &u(0) = 0 \qquad u(1) = 1 \end{aligned} \tag{96}$$

and

$$\begin{aligned} &\ddot{w} + \lambda q w = -g(t) \quad 0 \le t \le 1 \\ &w(0) = 0 \qquad w(1) = 0 \,. \end{aligned} \tag{97}$$

It is evident that the function x may be expressed in terms of the auxiliary functions u and w

$$x(t) = w(t) + cu(t) \tag{98}$$

w and u should be considered as functions of t and λ, and we write

$$w = w(\lambda, t) \quad \text{and} \quad u = u(\lambda, t) . \tag{99}$$

Differentiating Equation (96) with respect to λ, we obtain

$$\ddot{u}_\lambda + q(t)u + \lambda q(t)u_\lambda = 0 \tag{100}$$

with the boundary conditions

$$u_\lambda(0, \lambda) = 0 \quad \text{and} \quad u_\lambda(1, \lambda) = 0 . \tag{101}$$

The solution of (97) using the Green's function $G(t, y, \lambda)$ is

$$w(t, \lambda) = \int_0^1 G(t, y, \lambda)g(y)\, dy. \tag{102}$$

In terms of this Green's function the solution of the equation for u_λ, i.e., Equation (100) is

$$u_\lambda(t, \lambda) = \int_0^1 G(t, y, \lambda)q(y)u(y, \lambda)\, dy . \tag{103}$$

This is one of the basic differential equations with the initial condition

$$u(t, 0) = t \qquad 0 \le t \le 1 \tag{104}$$

as seen by inspection from (96). Similarly, from Equation (97), differentiating both sides with respect to λ, we have

$$\ddot{w}_\lambda + qw + \lambda q w_\lambda = 0 \tag{105}$$

with boundary conditions

$$w_\lambda(0, \lambda) = 0 \qquad w_\lambda(1, \lambda) = 0 . \tag{106}$$

The solution of (105) can be represented as

$$w_\lambda(t, \lambda) = \int_0^1 G(t, y, \lambda)q(y)w(y, \lambda)\, dy . \tag{107}$$

This is another basic equation with initial conditions at $\lambda = 0$ given by

$$w(t, 0) = \int_0^1 G(t, y, 0)g(y)\, dy . \tag{108}$$

The Green's function for $\lambda = 0$ is easily seen to be

$$G(t, y, 0) = \begin{cases} y(1-t) & 0 \le y \le t \\ t(1-y) & t \le y \le 1 \end{cases} \tag{109}$$

We have already seen from Equation (107) that

$$w_\lambda(t, \lambda) = \int_0^1 G(t, y, \lambda)q(y)w(y, \lambda)\, dy \tag{110}$$

and using Equation (102) we can write

$$w_\lambda(t, \lambda) = \int_0^1 G(t, y, \lambda)q(y)\, dy \int_0^1 G(y, y', \lambda)g(y')\, dy' . \tag{111}$$

From Equation (102), of course, we can express

$$w_\lambda(t, \lambda) = \int_0^1 G_\lambda(t, y, \lambda)g(y)\, dy . \tag{112}$$

Hence, we have, in view of the fact that g(y) is an arbitrary function, the differential equation for the Green's function given by

$$G_\lambda(t, y, \lambda) = \int_0^1 G(t, y', \lambda)q(y')G(y', y, \lambda)\, dy' \tag{113}$$

with initial conditions for G(t, y, 0) given by Equation (109). Hence, we have a Cauchy system for u, w and G, given by

$$u_\lambda(t, \lambda) = \int_0^1 G(t, y', \lambda)q(y')u(y', \lambda)\, dy' \tag{114}$$

$$w_\lambda(t, \lambda) = \int_0^1 G(t, y', \lambda)q(y')w(y', \lambda)\, dy' \tag{115}$$

and

$$G_\lambda(t, y, \lambda) = \int_0^1 G(t, y', \lambda)q(y')G(y', y, \lambda)\,dy' \qquad (116)$$

with the following set of initial conditions at $\lambda = 0$

$$u(t, 0) = t \qquad (117)$$

$$w(t, 0) = \int_0^1 G(t, y', 0)g(y')\,dy' \qquad (118)$$

and

$$G(t, y, 0) = \begin{cases} y(1-t) & 0 \le y \le t \qquad (119) \\ t(1-y) & t \le y \le 1\,. \qquad (120) \end{cases}$$

It can be shown by further analysis that u and w are the solutions of the boundary value problems indicated and that G is the Green's function, which is continuous in t and y with a jump discontinuity of G at $t = y$. For more details and other applications of this procedure to integro differential equations, see the original paper cited in references [10] and [11].

7. FREDHOLM RESOLVENT

We have seen in the previous chapter that we can write a quadratic functional associated with a Fredholm integral equation just as we did for differential equations. In the process of computing the minimum value of the functional employing dynamic programming methods, we arrive at interesting relations governing the resolvent operator corresponding to the Fredholm kernel. See Reference [12].

In order to indicate the type that can be expected, let us consider the linear integral equation

$$u(t) + v(t) + \int_a^T k(t, s)u(s)\,ds = 0 \quad 0 \le a \le T \qquad (121)$$

the 'Euler equation' associated with the minimum of the quadratic functional

$$J(u) = \int_a^T u^2(t)\,dt + 2\int_a^T u(t)v(t)\,dt + \int_a^T\int_a^T k(t, s)u(t)u(s)\,dt\,ds \tag{122}$$

over $u \in L^2(a, T)$.

We suppose that $k(t, s)$ is a symmetric kernel with the property that the quadratic term

$$\int_a^T\int_a^T k(t, s)u(t)u(s)\,dt\,ds + \int_a^T u^2(t)\,dt \tag{123}$$

is positive definite, i.e., bounded from below by $b\int_a^T u^2(t)\,dt$ for some $b > 0$ and $0 \leq a \leq T$. Then it is easy to demonstrate that the convex functional $J(u)$ possesses an absolute minimum furnished by the unique solution of (121). This solution may be represented in the form

$$u(t) = -v(t) + \int_a^T q(t, s, a)v(s)\,ds, \tag{124}$$

where the kernel $q(t, s, a)$ is called the Fredholm resolvent. Our objective is to obtain a nonlinear equation of Riccati type for $q(t, s, a)$.

A. Expression for min J(u)

We may write

$$J(u) = \int_a^T \left[u(t) + v(t) + \int_a^T k(t, s)u(s)\,ds\right]u(t)\,dt + \int_a^T u(t)v(t)\,dt \tag{125}$$

$$= \int_a^T u(t)v(t)\,dt$$

for the minimizing u. Using the expression for u in (124), we may write

$$\min_u J(u) = \int_a^T \left(-v(t) + \int_a^T q(t, s, a)v(s)\, ds\right)v(t)\, dt$$

$$= -\int_a^T v^2(t)\, dt + \int_a^T\int_a^T q(t, s, a)v(s)v(t)\, ds\, dt \qquad (126)$$

$$= f(v(t), a) .$$

Functional Equation for f(v, a). Let us write

$$J(u) = \int_a^{a+s} u^2(t)\, dt + 2\int_a^{a+s} u(t)v(t)\, dt + 2\int_a^{a+s}\int_{a+s}^T k(t, s)u(t)u(s)\, dt\, ds +$$

$$+ \int_{a+s}^T u^2(t)\, dt + \int_{a+s}^T u(t)v(t)\, dt + \int_{a+s}^T\int_{a+s}^T k(t, s)u(t)u(s)\, dt\, ds + \qquad (127)$$

$$+ \int_a^{a+s}\int_a^{a+s} k(t, s)u(t)u(s)\, dt\, ds .$$

Regarding s as an infinitesimal, it follows in the usual fashion that

$$f(v, a) = \min_u J(u) = \min_u [s[u^2(a) + 2u(a)v(a)] + f(v(t) + su(a)k(a, t), a+s)] + 0(s) . \qquad (128)$$

Since f(v, a) is clearly a differentiable function of v(x) and a, we may obtain a differential equation for f(v, a). To this end let us define

$$L(w(t), a) = \lim_{s\to 0}\left[\frac{f(v(t) + sw(t), a) - f(v(t), a)}{s}\right]. \qquad (129)$$

Then the limiting form of (128) is

$$0 = \min_{u(a)}\left[u^2(a) + 2u(a)v(a) + u(a)L[k(a, t), a] + \frac{\partial f}{\partial a}\right]. \qquad (130)$$

The minimum is attained at

$$u(a) = -v(a) - \frac{L(k(a,\ t),\ a)}{2} \tag{131}$$

yielding the quadratically nonlinear equation

$$\frac{\partial f}{\partial a} = \left[v(a) + \left\{\frac{L(k(a,\ t),\ a)}{2}\right\}\right]^2 . \tag{132}$$

b. The Form of L(k(a, t), a

From the expression for f(v, a), we readily see that

$$L(w,\ a) = -2\int_a^T v(t)w(t)\ dt + 2\int_a^T\int_a^T q(t,\ s,\ a)v(t)w(s)\ dt\ ds\ . \tag{133}$$

Thus,

$$L(k(t,\ a),\ a) = -2\int_a^T v(t)k(a,\ t)\ dt + 2\int_a^T\int_a^T q(t,\ s,\ a)k(a,\ s)v(t)\,dt\,ds$$
$$= \int_a^T M(k)v(t)\ dt, \tag{134}$$

where $M(k) = M(k,\ t)$.

Functional Equation for q(t, s, a). Using this expression in (132) we obtain the relation

$$\frac{\partial f}{\partial a} = \left(v(a) + \int_a^T \frac{M(k)v(t)\ dt}{2}\right)^2$$
$$= v^2(a) + v(a)\int_a^T M(k)v(t)\ dt + \int_a^T\int_a^T \frac{M(k,\ t_1)M(k,\ t_2)v(t_1)v(t_2)\ dt_1\ dt_2}{4} . \tag{135}$$

On the other hand, using (126)

$$\frac{\partial f}{\partial a} = v^2(a) - 2v(a)\int_a^T q(a,\ s,\ a)v(s)\,ds + \int_a^T\int_a^T \frac{\partial q}{\partial a}(t,\ s,a)v(t)v(s)\ \ dt\,ds. \tag{136}$$

Comparing (134), (135) and (136), which holds for all $v \in L^2(a, T)$, we obtain the two relations

$$q(a, t; a) = k(a, t) - \int_a^T q(t, w, a)k(a, w)\, dw \qquad (137)$$

$$\frac{\partial q}{\partial a}(t, s, a) = \frac{M(k, t)M(k, s)}{4} = q(a, t, a)q(a, s, a). \qquad (138)$$

The last relation is the desired analog of the Riccati equation in the finite-dimensional case of ordinary differential equations. This can also be obtained in a straightforward way obtaining the integral equation for $q(x, y, t)$ with $k(x, y)$ as the inhomogenous term by substituting Equation (124) into Equation (121).

8. THE RICCATI EQUATION

A first order nonlinear differential equation given by

$$\frac{du}{dt} = p_1(t) + p_2(t)u + p_3(t)u^2; \quad u(0) = c \qquad (139)$$

is called the Riccati equation and plays an important role in many fields, as we have seen, in dynamic programming, control theory, mathematical physics and other field of engineering, etc. A general way of aaproaching this equation is to note that the linear equation

$$v'' + a_1 v' + a_2 v = 0, \qquad (140)$$

where a_1 and a_2 are functions of t goes over to the equation of the form (139) upon change of variable

$$v = \exp \int_0^t u\, dt_1 \qquad (141)$$

so that we have

$$\begin{aligned} v' &= uv \\ v'' &= (u'v + uv') = (u' + u^2)v . \end{aligned} \qquad (142)$$

Hence, if $v \neq 0$ which is an important condition, we obtain

$$u' + u^2 + a_1 u + a_2 = 0 \ . \tag{143}$$

Conversely, if we set $u = v'/v$ in the above equation, we obtain the linear equation of second order for v.

If v satisfies the Riccati equation

$$v' + v^2 + pv + q = 0 \tag{144}$$

it can be easily seen that v^{-1} also satisfies the same type of Riccati equation

$$\begin{aligned} v^{-1} &= \bar{v} \\ \bar{v}' &= 1 + p\bar{v} + q\bar{v}^2 \ . \end{aligned} \tag{145}$$

It can also be shown that w, obtained from v, by the transformation

$$w = \frac{av + b}{cv + d} \tag{146}$$

satisfies a Riccati type equation.

9. QUASILINEARIZATION

We have referred to this method in previous chapters and the aim in developing this method is to obtain representation of the solutions of nonlinear equations under suitable conditions in terms of a maximum or minimum operation. This, when applied to Riccati equations, leads, under suitable circumstances, to obtaining a method for successive approximations to the solutions which possess the properties of monotonicity and quadratic convergence.

Starting with the Riccati Equation (144)

$$v' = -v^2 - p(t)v - q(t) \tag{147}$$

let us replace v^2 by its equivalent expression $\max_u [2uv - u^2]$. The equation now has the form

$$v' = -\max_u [2uv - u^2] - p(t)v - q(t) \tag{148}$$

$$= \min_u [u^2 - 2uv - pv - q(t)] \ . \tag{149}$$

Equation (149) is linear but for the minimization operation. Hence, the name 'quasilinearization'. Consider the comparison equation

$$w' = u^2 - 2uw - pw - q , \tag{150}$$

where u is a fixed function of t. Let v and w have the same initial values $w(0) = v(0) = c$. Let us assume that the solution $v(t)$ exists in the interval $(0, t)$. We now establish that $w \geq v$ by observing that

$$v' \leq u^2 - 2uv - pv - q(t) \tag{151}$$

for all $u(t)$.

$$v' = u^2 - 2uv - pv - q - r(t) , \tag{152}$$

where $r(t) \geq 0$ for $t \geq 0$. The function $r(t)$ depends, of course, on u and v.

Let us now digress a little to consider a linear equation of the form

$$\frac{dw}{dt} - f(t)w = g(t); \quad w(0) = c . \tag{153}$$

The solution of this equation can easily be found as

$$\begin{aligned} w &= c \exp \int_0^t f(s)\,ds + e^{\int_0^t f(s)\,ds} \int_0^t g(s) e^{-\int_0^s f(r)\,dr} \\ &= T(f, g) \end{aligned} \tag{154}$$

which defines the operator T, a linear inhomogeneous operator on g. The positivity of the operator T is evident from (154). Hence, if $g_1(t) \geq g_2(t)$ for $t \geq 0$, $T(f, g_1) \geq T(f, g_2)$ for $t \geq 0$ which is an important property.

Let us now go back to Equation (152) and write v as

$$v = T[-2u - p(t), u^2 - q(t) - r(t)] \tag{155}$$

$$\leq T[-2u - p(t), u^2 - q(t)] = w . \tag{156}$$

Since this inequality holds for all $u(t)$, with equality for $u(t) = v(t)$, we have the important result

$$v(t) = \min_u T[-2u - p(t),\ u^2 - q(t)]\ . \tag{157}$$

Hence, we have the representation for $v(t)$ as

$$\begin{aligned} v(t) = \min_u \Big\{ c \exp\Big[-\int_0^t [2u(s) + p(s)]\, ds\Big] &+ \\ + \int_0^t \exp\Big[-\int_s^t [2u(r) + p(r)]dr\Big]\Big[u^2(s) - q(s)\Big]\, ds\Big\}. \end{aligned} \tag{158}$$

Thus using quasilinearization we have obtained the analytic expression for the solution of the Riccati equation.

From this result we generate a series of approximations to the solutions of $v(t)$. We know that the minimum is attained by the solution $v(t)$ itself. Hence, we suspect that if v_0 is a reasonable initial approximation, then v_1 obtained as the solution of

$$v_1' = v_0^2 - 2v_0v_1 - p(t)v_1 - q(t); \quad v_1(0) = c \tag{159}$$

will be even a better approximation, by analogy with the procedure employed to find the roots of $f(x) = 0$ using the Newton-Raphson technique. Repeating this we obtain the recurrence relations

$$v_{n+1}' = v_n^2 - 2v_nv_{n+1} - p(t)v_{n+1} - q(t); \quad v_{n+1}(0) = c. \tag{160}$$

Generally, if the differential equation were $v' = g(v, t)$, we would use the approximation scheme

$$v_{n+1}' = g(v_n, t)+(v_{n+1}-v_n)g_v(v_n, t); \quad v_{n+1}(0) = c\ . \tag{161}$$

The scheme may also be called the Newton-Raphson-Kantorovich scheme. In the special case when $g(v, t) = -v^2 - p(t)v - q$ Equation (161) reduces to Equation (160). From the geometric background of the Newton-Raphson approximation procedure, we expect quadratic convergence. To establish the monotonicity of the approximation scheme let us start with the recurrence relation

$$v'_{n+1} = v_n^2 - 2v_n v_{n+1} - p(t)v_{n+1} - q; \quad v_{n+1}(0) = c \,. \qquad (162)$$

Since the minimum value of the right-hand side is obtained for $v_n = v_{n+1}$, we have

$$v'_{n+1} = v_n^2 - 2v_n v_{n+1} - pv_{n+1} - q \geq v_{n+1}^2 - 2v_{n+1}v_{n+1} - pv_{n+1} - q. \quad (163)$$

We now compare this equation with the relation for v_{n+2}

$$v'_{n+2} = v_{n+1}^2 - 2v_{n+1}v_{n+2} - pv_{n+2} - q; \quad v_{n+2}(0) = c \,. \quad (164)$$

Hence, using the property of the linear operator T defined in Equation (154), we find that

$$w_{n+1} \geq v_{n+2} \,. \qquad (165)$$

Therefore, inductively we see that

$$v_1 \geq v_2 \geq v_3 \cdots \,. \qquad (166)$$

If the function v_0 is chosen arbitrarily it need not be true that $v_0 \geq v_1$. However, one can always show that $v_n(t) \geq v(t)$ for all $n \geq 1$ since we can write

$$v' = -v^2 - pv - q \qquad (167)$$

$$= \min_w [w^2 - 2wv - pv - q] \leq v_n^2 - 2v_n v - pv - q \,. \qquad (168)$$

Therefore, it is evident that $v \leq v_n$ for each $n \geq 1$. In any interval [0, t] within which v(t) exists, we can assert the convergence of $\{v_n\}$. From the relation

$$v_{n+1}(t) = \int_0^t [v_n^2 - 2v_n v_{n+1} - pv_{n+1} - q] \, dt_1 + C \qquad (169)$$

we conclude on the basis of bounded convergence that

$$\bar{v} - c = \int_0^t [-\bar{v}^2 - p\bar{v} - q]\, dt \;, \tag{170}$$

where $\bar{v} = \lim_{n\to\infty} v_n(t)$. Hence, $\bar{v}(t)$ as an integral, is continuous and differentiable. We obtain therefore

$$\bar{v}' = - \bar{v}^2 - p\bar{v} - q; \quad \bar{v}(0) = c \;. \tag{171}$$

The uniqueness theorem shows that $v = \bar{v}$. Since the elements of $\{v_n\}$ are continuous, since convergence is monotone and since $v(t)$ is continuous, we can assert using Dini's theorem that convergence is actually uniform in $[0, t]$.

In the above treatment, we have obtained approximations $v_n (n \geq 1)$ which are upper bounds to the actual solution v. In some other types of equations we may be obtaining successive approximations which give lower bounds to the solution. As an illustration, we begin with such an equation

$$\frac{du}{dt} = u^2 + a(t); \; u(0) = c \tag{172}$$

which, using quasilinearization, may be written in the form

$$\frac{du}{dt} = \max_v \; [2uv - v^2 + a(t)] \quad u(0) = c \;. \tag{173}$$

This implies that for any function v

$$\frac{du}{dt} \geq 2uv - v^2 + a(t); \quad u(0) = c \tag{174}$$

with an initial approximation for u as u_0, substituted for v, we obtain

$$u_1' = 2u_0 u_1 - u_0^2 + a(t); \quad u_1(0) = c \;. \tag{175}$$

Then we obtain the sequence of functions $\{u_n\}$ having the recurrence relation

$$u_{n+1}' = 2u_{n+1}u_n - u_n^2 + a(t); \quad u_{n+1}(0) = c \;. \tag{176}$$

As demonstrated in the earlier case, we can show the monotonicity of $\{u_n\}$ sequence to be

$$u_1 \le u_2 \le \cdots u_{n+1} \le \cdots \tag{177}$$

and also

$$u_n \le u(t) \quad \text{for} \quad n \ge 1 \ . \tag{178}$$

To show the quadratic convergence of the sequence we write Euqation (176) as

$$u_n' = u_{n-1}^2 + 2(u_n - u_{n-1})u_{n-1} + a(t); \quad u_n(0) = c \tag{179}$$

and

$$u' = u_{n-1}^2 + 2(u-u_{n-1})u_{n-1} + (u-u_{n-1})^2 + a(t); \quad u(0) = c \ . \tag{180}$$

The second equation is a consequence of the identity

$$u^2 = (u_{n-1}+u-u_{n-1})^2 = u_{n-1}^2 + 2u_{n-1}(u-u_{n-1}) + (u-u_{n-1})^2 . \tag{181}$$

It follows

$$\begin{aligned} &(u-u_n)' = 2(u-u_n)u_{n-1} + (u-u_{n-1})^2 \\ &\text{and} \quad (u-u_n)(0) = 0 \cdots \end{aligned} \tag{182}$$

Integrating this with $(u-u_{n-1})^2$ as a forcing term

$$(u-u_n) = \exp\left(2\int_0^t u_{n-1}\, ds\right)\int_0^t \exp\left(-2\int_0^{t_1} u_{n-1}\, dr\right)(u-u_{n-1})^2\, dt_1 \ . \tag{183}$$

Thus

$$\begin{aligned} |(u-u_n)| &\le \int_0^t \left[\exp 2\int_{t_1}^t u\, ds\right](u-u_{n-1})^2\, dt_1 \\ &\le \int_0^t e^{2k(t-t_1)}\, dt_1 \left[\max_t |u-u_{n-1}|\right]^2 , \end{aligned} \tag{184}$$

where $k = \max_t |u|$. Thus we obtain the inequality

$$\max_t |u-u_n| \leq \max_t \int_0^t e^{2k(t-t_1)}\, dt_1 \left[\max_t |u-u_{n-1}|^2\right]$$
$$\leq (\text{constant}) \max_t |u-u_{n-1}|^2 . \tag{185}$$

This is the type of quadratic convergence that we obtain for the sequence $\{u_n\}$ by the method described above. Thus we have seen how quasilinearization techniques lead to finding successively better and better approximations to the solution. We can get either the lower bound or the upper bound as the case may be depending on the structure of the equation, as we have seen in the two cases. To get both upper and lower bounds for the solution of the same equation, a transformation of the dependent variable such as writing down the equation for its inverse may help in some cases (References [13], [14], and [15]).

We know from the previous chapters that a multidimensional transport process in a one-dimensional medium leads to a set of coupled equations for the up and down streams and that the reflection matrix satisfies the matrix Riccati equation.

We can arrive at the Riccati equation by taking a linear system of matrix equations with suitable boundary values.

$$\begin{aligned} x' &= Ax + By \\ & \qquad\qquad 0 \leq t \leq d \\ y' &= Cx + Dy \end{aligned} \tag{186}^*$$

and making the transformation $z = xy^{-1}$. Let the inverse y^{-1} exist in some interval $[0, t]$. Then

$$\begin{aligned} \frac{dz}{dt} &= \left(\frac{dx}{dt}\right)y^{-1} - xy^{-1}\left(\frac{dy}{dt}\right)y^{-1} \\ &= (Ax + By)y^{-1} - xy^{-1}[Cx + Ly]y^{-1} \\ &= Az + B - zCz - zD . \end{aligned} \tag{187}$$

Conversely we can transform the equation for z into a system of two equations. We can also show that z^{-1} and $(A_1z + B_1)(C_1z + D_1)^{-1}$ satisfy Riccati equations of the form (187).

---

* x, y and z are to be understood as vectors as the context indicates, since A, B, C and D are matrices for a many particle system.

Let us now deal with a simple Riccati equation and see whether the methods employed in the scalar case do apply. Let us consider the equation

$$R' = A - R^2; \qquad R(0) = I , \tag{188}$$

where A is positive definite (A is symmetric).

Before going further let us clarify our notions of inequalities in the case of matrices. Here we deal with real matrices, and when we say $A \geq B$, A and B being symmetric, to signify (A-B) is nonnegative definite. This is equivalent to the scalar inequality $(x, Ax) \geq (x, Bx)$ for all real vectors x. This is only partial ordering since $A \geq B$ does not necessarily imply $A^2 \geq B^2$. However, an important result is that $A \geq B > 0$ implies that

$$A^{-1} \leq B^{-1} . \tag{189}$$

To prove this we consider the representation of the inverse of a matrix A by using the quasilinearization technique as

$$-(y, A^{-1}y) = \min_x [(x, Ax) - 2(x, y)] \tag{190}$$

for $A > 0$. The minimum occurs as is evident when $Ax = y$. Hence, it is obvious that $A^{-1} \leq B^{-1}$.

Now let us go back to Equation (188). We can express R^2 as

$$R^2 = (S+R-S)^2 = S^2 + S(R-S) + (R-S)S + (R-S)^2 , \tag{191}$$

where R and S are real symmetric matrices. Hence,

$$\begin{aligned} R^2 &\geq S^2 + S(R-S) + (R-S)S \\ &\geq SR + RS - S^2 . \end{aligned} \tag{192}$$

Since we write Equation (188) as

$$R' \leq (A+S^2) - (SR+RS), \quad R(0) = I . \tag{193}$$

Consider the associated equation

$$W' = (A+S^2) - (SW+WS); \quad W(0) = I . \tag{194}$$

The solution of this equation can be easily found as

$$W = y^{-1}z^{-1} + \int_0^t y^{-1}(t)y(t_1)[A+S^2]z(t_1)z^{-1}(t)\,dt_1 , \qquad (195)$$

where y and z are non-singular matrices governed by the equations

$$y' = +y(t)S; \quad y(0) = I \qquad (196)$$

$$z' = S(t)z; \quad z(0) = I . \qquad (197)$$

Hence, z is the transpose of y and $y^{-1}(t_1)y(t)$ is the transpose of $z(t)z^{-1}(t_1)$ and W is symmetric for $t \geq 0$. It follows if we regard Equation (193) to be of the form

$$R' = (A+S^2) - (SR+RS) - p(t); \quad R(0) = I , \qquad (198)$$

where $p \geq 0$ for $t \geq 0$ we can write

$$R = W - \int_0^t y^{-1}(t)y(t_1)p(t_1)z^{-1}(t)\,dt_1 . \qquad (199)$$

The second integral is clearly non-negative for $t \geq 0$. Hence, we have the inequality

$$R \leq W(S, t) . \qquad (200)$$

We call this solution W(S, t) as $W_u(S, t)$ since this is an upper bound to R. Next we make the change of variable.

$R = M^{-1}$. The equation for M is then obtained from Equation (188)

$$M' = I - MAM . \qquad (201)$$

We start with the identity

$$\begin{aligned} MAM &= (S+M-S)A(S+M-S) \\ &= SAS + SA(M-S) + (M-S)AS + (M-S)A(M-S) \\ &\geq SAS + SA(M-S) + (M-S)AS \\ &\geq SAM + MAS - SAS \end{aligned} \qquad (202)$$

for all symmetric S. Therefore,

$$M' \leq I + SAS - (SAM + MAS) . \tag{203}$$

The solution of the associated equation

$$W' = I + SAS - (SAW + WAS) \tag{204}$$

can be found and following the methods advanced earlier can be shown to be the lower bound $W_L(S, t)$. Hence, for any two symmetric matrices suitably chosen we can set the bounds for R. (See Reference [16].)

$$W_L(S_1, t) \leq R(t) \leq W_U(S_2, t) . \tag{205}$$

10. THE CROSS-RATIO RELATIONS

As we very well know the set of complete homogeneous equations

$$\begin{aligned} u'(t) - au - bv &= L_1(u, v) = 0 \\ -v'(t) - dv - cu &= L_2(u, v) = 0 \end{aligned} \tag{206}$$

can be associated with the scalar Riccati equation

$$L(r) = r' + (a+d)r + br^2 + c = 0 , \tag{207}$$

where a, b, c and d are assumed continuous on the real interval I. If there exist solutions of u and v of the set (206) in the interval I_0 of I such that $u \neq 0$ for $t \in I_0$, then r(t) the solution of (207) is given by $r(t) = v/u$ for $t \in I_0$. If u and v are continuously differentiable, one may easily check the identity

$$u^2L(r) = -vL_1(u, v) - uL_2(u, v) . \tag{208}$$

If r(t) is known, u and v can be computed by simple quadrature.

Let us take a solution r_0 which is a solution of the Riccati Equation (207) with certain initial value in the interval I_0 of I. Another solution r_1 of the same equation I_0 can be related to r_0 by the relation

$$r_1(t) = r_0(t) + \frac{g(t/r_0)h(t/r_0)k}{1+kf(t/r_0)} , \tag{209}$$

where g, h and f are defined as follows

$$g(t \mid r_0) = \exp\left\{-\int_0^t [d(\tau) + r_0(\tau)b(\tau)]\, d\tau\right\} \tag{210}$$

$$h(t \mid r_0) = \exp\left\{-\int_0^t [a(\tau) + b(\tau)r_0(\tau)]\, d\tau\right\} \tag{211}$$

$$f(t \mid r_0) = \int_0^t g(\tau \mid r_0)b(\tau)h(\tau \mid r_0)\, d\tau \tag{212}$$

and k is a constant $= r_1(0) - r_0(0)$. (213)

Also the essential condition that should hold is that $1 + kf(t \mid r_0) \neq 0$ for $t \in I_0$. It is easy to verify by actually differentiating (209) and using Equation (210) through (211) so that $r_1(t)$ satisfies the same Riccati Equation (207). Since $g(0) = h(0) = 1$ and $f(0) = 0$, it is also readily seen from (209) that k the constant is given by Equation (213). Hence, if we have the solution r_1 and r_2 which are not equal at any point in the interval, then since

$$r_1 - r_0 = \frac{g(t \mid r_0)h(t \mid r_0)k_1}{1 + k_1 f(t \mid r_0)} ; \quad k_1 = r_1(0) - r_0(0) \tag{214}$$

and

$$r_2 - r_0 = \frac{g(t \mid r_0)h(t \mid r_0)k_2}{1 + k_2 f(t \mid r_0)} ; \quad k_2 = r_2(0) - r_0(0) \tag{215}$$

we conclude

$$r_1 - r_2 = \frac{g(t \mid r_0)h(t \mid r_0)(k_1-k_2)}{[1+k_1 f(t \mid r_0)][1+k_2 f(t \mid r_0]} . \tag{216}$$

If we have four independent solutions r_1, r_2, r_3, and r_4 on the interval I_0, then their cross ratio is a constant

$$\frac{r_3-r_1}{r_3-r_2} \cdot \frac{r_4-r_2}{r_4-r_1} = \frac{k_3-k_1}{k_3-k_2} \cdot \frac{k_4-k_2}{k_4-k_1} \quad . \tag{217}$$

These relations may prove to be of importance in many practical applications. (See References [17], [18], and [19].)

11. MATRIX RICCATI EQUATION AND AUXILIARY FUNCTIONS

Let us consider the coupled set of vector matrix equations given by

$$\frac{dU}{dt} = AU + BV \quad a \leq t \leq T \tag{218}$$

$$-\frac{dV}{dt} = CU + DV. \tag{219}$$

U and V being column vectors of order n, A, B, C, and D matrices are either square matrices of order n × n or suitable dimensions. The solution to the above equation can be given by the state transition matrix

$$U(t) = \varphi_{11}(t, a)U(a) + \varphi_{12}(t, a)V(a) \tag{220}$$

$$V(t) = \varphi_{21}(t, a)U(a) + \varphi_{22}(t, a)V(a) , \tag{221}$$

where φ_{ij} are the partitioned state transition matrices satisfying

$$\frac{d}{dt}\begin{bmatrix} \varphi_{11} & \varphi_{12} \\ \varphi_{21} & \varphi_{22} \end{bmatrix} = \begin{bmatrix} A(t) & B(t) \\ -C(t) & -D(t) \end{bmatrix}\begin{bmatrix} \varphi_{11} & \varphi_{12} \\ \varphi_{21} & \varphi_{22} \end{bmatrix} \tag{222}$$

with the initial condition

$$\varphi_{ij}(a, a) = \begin{cases} 1 & i = j \\ 0 & i \neq j \end{cases} . \tag{223}$$

To derive the Riccati equations associated with the system, let us assume that U(t) and V(t) can be defined by the equations

$$U(t) = P_{11}(t, a)U(a) + P_{12}(t, a)V(t) \tag{224}$$

$$V(a) = P_{21}(t, a)U(a) + P_{22}(t, a)V(t) . \tag{225}$$

It is a simple exercise in matrix algebra to rearrange the functions in the system of Equations (224), (225); (222) and (220), (221) to obtain the P_{ij}'s in terms of φ_{ij}'s

$$P_{11} = \varphi_{11}(t, a) - \varphi_{12}(t, a)\varphi_{22}^{-1}(t, a)\varphi_{21}(t, a) \tag{226}$$

$$P_{12} = \varphi_{12}(t, a)\varphi_{22}^{-1}(t, a) \tag{227}$$

$$P_{21} = -\varphi_{22}^{-1}(t, a)\varphi_{21}(t, a) \tag{228}$$

$$P_{22} = \varphi_{22}^{-1}(t, a) , \tag{229}$$

where

$$P_{ij}(a, a) = \begin{cases} I & i = j \\ 0 & i \neq j . \end{cases} \tag{230}$$

The differential equations for P_{ij}'s are easily obtained in view of the equations satisfied by φ_{ij}'s. We have

$$\frac{dP_{12}}{dt} = B + AP_{12} + P_{12}D + P_{12}CP_{12} \tag{231}$$

$$\frac{dP_{11}}{dt} = [A + P_{12}C]P_{11} \tag{232}$$

$$\frac{dP_{22}}{dt} = P_{22}\ [D+CP_{12}] \tag{233}$$

$$\frac{dP_{21}}{dt} = P_{22}CP_{11} . \tag{234}$$

Equation (231) is the Riccati equation for P_{12}, with the initial condition $P_{12}(a, a) = 0$. The other three auxiliary functions depend on the solution of the first Riccati equation. Thus, we find that the coefficients of any Riccati equation completely determine the state-equation (222) as well as the auxiliary equation. Hence, if the state transition matrix of the system is determined, the Riccati equation can be solved. However, the initial condition for the Riccati equation which we set out to solve may not have the initial condition $P_{12}(a, a) = 0$. If the initial condition is zero, the solution of the state-stransition matrix leads to the solution of the Riccati equation. For arbitrary initial conditions the auxiliary functions are necessary to compute the solution as we shall see below. Let the initial value of the Riccati equation (231) be

$$\bar{P}_{12}(a, a) = RI . \tag{235}$$

Let us now define a new state transition matrix, $\psi(t, a)$ related to φ in the following manner:

$$\psi(t, a) = \begin{bmatrix} \varphi_{11}(t, a) & \varphi_{12}(t, a) \\ \varphi_{21}(t, a) & \varphi_{22}(t, a) \end{bmatrix} \begin{bmatrix} I & RI \\ 0 & I \end{bmatrix} \tag{236}$$

$\psi(t, a)$ will satisfy the same equation as φ with a different initial condition and the solution $\bar{P}_{12}(t, a)$ will now be given by

$$\begin{aligned} \bar{P}_{12}(t, a) &= \psi_{12}(t, a)\psi_{22}^{-1}(t, a) = \\ &= [\varphi_{11}(t, a)RI + \varphi_{12}(t, a)][\varphi_{21}(t, a)RI+\varphi_{22}(t, a)]^{-1}. \end{aligned} \tag{237}$$

We can eliminate the φ's and express $\bar{P}_{12}$ in terms of the old P_{ij}'s as

$$\bar{P}_{12}(t, a) = P_{12}(t, a)+P_{11}(t, a)[1-RIP_{21}(t, a)]^{-1}RIP_{22}(t, a) . \tag{238}$$

This $\bar{P}_{12}$ satisfies the Riccati equation and the initial conditions. If the state transition matrix ψ is found the solution to the Riccati equation and the auxiliary functions can be computed with arbitrary initial values. Recursive algorithms have been used to compute φ matrices, using their semigroup properties.

$$\varphi(t + \Delta, a) = \varphi(t + \Delta, t)\varphi(t, a) . \tag{239}$$

For details of these discussions see Denman and Rao [20]. Let us now take four independent solutions of the Riccati Equation (231) with different initial conditions, i.e., with $R_i I \neq R_j I$ and cross ratio of the solutions as we did in the scalar case. Let $\bar{P}_{21}(a, a) = R_i I$ and let $P_{21}(r, a) = Y_i$ for i = 1, 2, 3 and 4. Then, if we compute the cross ratio matrix S(t)

$$S(t) = (Y_1-Y_3)(Y_3-Y_2)^{-1}(Y_2-Y_4)(Y_4-Y_1)^{-1} \tag{240}$$

using the expression (238), we find

$$S(t) = P_{11}(t, a)[I-RI_1P_{21}(t, a)]^{-1}[RI_1-RI_3][RI_3-RI_2]^{-1}$$
$$[RI_2-RI_4][RI_4-RI_1]^{-1}[I-RI_1P_{21}(t, a)]P_{11}^{-1}(t, a) . \tag{241}$$

This is t dependent and the value of S(t) varies with t. For the scalar case this turns out to be the usual result. The method of computing the solution of the Riccati equation using the auxiliary functions has found applications in many engineering problems. Numerically computing the state transition matrix is very helpful in avoiding some well-known difficulties in computing the solutions of the Riccati equation such as stepping over the singular points (see Reference [20]).

Thus we have briefly summarized different ways of looking at the solutions of the second order equations with boundary conditions. Different types of analyses may suit different situations.

REFERENCES

[1] Bellman, R., 'On Variation Diminishing Properties of Green's Functions', Boll. D'Unione Math. 16 (1961), 164.

[2] Bellman, R., 'Oscillatory and Unimodal Properties of Solutions of Second Order Differential Equations', Bull. U.M.I. 10 (1964), 306.

[3] Bellman, R. and S. Lehman, 'Functional Equations in the Theory of Dynamic Programming - X: Resolvents Characteristic Functions and Values', Duke Mathematical Journal 27 (1960), 55.

[4] Miller, K.S. and M.M. Schiffer, 'On the Green's Functions of Ordinary Differential Equations', Proceedings of the American Mathematical Society 3 (1952), 433.

[5] Bellman, R. and H. Osborn, 'Dynamic Programming and the Variation of Green's Functions', Journal of Mathematics and Mechanics 7 (1958), 81.

[6] Bellman, R., 'On the Determination of Characteristic Values for a Class of Sturm-Liouville Problems', Illinois Journal of Mathematics 2 (1958), 577.

[7] Collatz, L., Numerische Behandlung von Differentialgleichungen, Springer-Verlag, Berlin, 1951.

[8] Shoemaker, C., 'Computation of Characteristic Values of Sturm-Liouville Problems with a Digital Computer', USC Technical Report No. USCEE-267, Department of Electrical Engineering, University of Southern California, Los Angeles, April 1968.

[9] Huss, R., H. Kagiwada, and R. Kalaba, 'A Cauchy System for the Green's Function and the Solution of a Two-Point Boundary Value Problem', Technical Report No. RB70-17, Department of Electrical Engineering, University of Southern California, April 1970; J. Franklin Institute 291 (1971), 159.

[10] Kalaba, R., Journal of Mathematical Physics 11 (1970) 1999.
Huss, R. and R. Kalaba, Journal Opt. The. Appl. 6 (1970) 415.

[11] Huss, R., R. Kalaba, and R. Vasudevan, 'On Boundary Value Problems for Integro Differential Equations', Journal of Mathematical Physics 15 (1974), 1285.

[12] Bellman, R., 'Functional Equations in the Theory of Dynamic Programming - VII: A Partial Differential Equation for the Fredholm Resolvent', Proceedings of the American Mathematical Society 8 (1957), 453.

[13] Bellman, R. and R. Kalaba, Quasilinearization and Nonlinear Boundary Value Problems, American Elsevier Publishing Co., Inc., New York, 1965.

[14] Bellman, R., 'Functional Equations in the Theory of Dynamic Programming - V: Positivity and Quasi-Linearity', Proceedings of the National Academy of Sciences 41 (1955), 743-746.

[15] Kalaba, R., 'On Nonlinear Differential Equations: The Maximum Operation and Monotone Convergence', Journal of Mathematics and Mechanics8 (1959), 519.

[16] Bellman, R., Methods of Nonlinear Analysis, Vol. II, Academic Press, Inc., 1978, New York.

[17] Reid, W.T., Riccati Differential Equations, Academic Press, Inc., New York, 1972.

[18] Fymat, A.L. and R. Vasudevan, 'A New Approach to Radiative Transfer Using Jone's Vector', Astrophysics and Space Science 38 (1975), 95.

[19] Davis, H.T., Introduction to Nonlinear Differential and Integral Equations, Dover Publications, Inc., New York 1962.

[20] Denman, E.D. and H.S. Rao, (a) 'On the Matrix Riccati Differential Equation' (to be published); (b) 'A Forward Integration Algorithm for the Linear Regulator Problem', Proceedings of the IEEE Conference, Dallas, Texas, 1970 (SWIEEECO).

CHAPTER XII

APPROXIMATIONS

1. INTRODUCTION

The numerical solutions of partial differential equations play a prominent role in numerical weather forecasting, optimal control theory, radiative transfer, and in many other areas of physics, engineering, biology and hydrodynamics. Solutions for such equations can be obtained by techniques using difference methods on a digital computer. These methods lead to the theory of stability of the solutions. Due to lack of space, we direct the readers to the references cited as [21], [22]. The standard finite difference methods require large computing time and storage. We will turn our attention to other methods like 'Differential quadrature' (Reference [1], see also Reference [6] and the spline approximation(Reference [2])). These are described in Sections 2 and 3.

In many practical situations we may be satisfied with approximate solutions. If we are trying to solve a wave equation of the type

$$u'' + f(x)u = 0$$

and if f(x) can be approximated in various regions by straight lines or polynomials or other known functions, then the solution of the above is obtained analytically in each region. Then we are in possession of an approximate solution to the wave equation. This idea of segmental approximation is explained in Sectins 4 and 5. The last section derives the spline approximation employing the methods of dynamic programming (see Reference [3]).

2. QUADRATURE

The type of approximation of the form

$$Lu = \sum_{i=1}^{N} a_i u(x_i) \tag{1}$$

is called 'quadrature' approximation when the operator L is an integral operator. A quadrature technique is a numerical procedure for the evaluation of a definite integral in terms of a linear sum of the values of the integrand at specific points. A typical formula in the one-dimensional case assumes the form

$$\int_0^1 f(t)\,dt = \Sigma\, W_i f(r_i) \ , \qquad (2)$$

r_i are the points at which f(r) has to be evaluated, W_i are the weights. The details of this can be found in Reference [4] (this reference lists relevant literature).

In analogy, when L is a linear differential operator, the above approximation is named 'differential quadrature' by Bellman [1]. In either case, the coefficients a_i of Equation (1) can be determined using interpolating polynomials. Although integral quadratures with a variety of interpolating polynomials are fully developed, the differential quadratures are still in an early stage of development. Differential quadrature has obvious applications in the numerical solution of partial differential equations, ordinary differential equations with boundary conditions and identification problems. In this chapter, we deal mostly with the former case.

In general, the interpolation formulas are expressed in two ways, either in terms of differences of the function or in terms of values of the function. The derivatives of a function can also be expressed in the same manner in both cases. The first case, i.e. the approximation of derivatives with differences, provides the basis of many methods of solving differential equations (the difference methods) as pointed out in the earlier chapters. This approxiamtion can also be used for the estimation of errors in numerical differentiation.

In the differential quadrature method we use the second approximation for solution of partial differential equations, and effectively smooth the 'data' by employing spline interpolation for determination of coefficient a_i of Equation (1).

The differential quadrature method using spline approximation is an efficient, quick, simple and powerful technique for different types of differential equations. It is expecially useful in cases where moderately accurate results, at few points of interest, using an on-line mini-computer, are required.

An advantage of the method of differential quadrature (over the conventional finite difference methods) is that the method allows us to change the structure of the model equations according to the different hypotheses (physiological, biological and medical) with extreme ease and care. This, in turn, afford experimenters in various fields an interactive real time testing mode between the hypothesis and experiment.

3. DIFFERENTIAL QUADRATURE

For simplicity, let us consider a first order partial differential equation, linear or nonlinear, of the form

$$u_t = g(t, x, u, u_x); \quad t \geq 0 \tag{3}$$

with any given general boundary condition on a finite interval $a \leq x \leq b$. If the function u, satisfying Equation (3) is sufficiently smooth, one can write the approximate relation

$$u_x(t, x_j) = \sum_{i=1}^{N} u(t, x_i)a_{ij}, \quad j = 1, \cdots, N . \tag{4}$$

Substitution of the above differential quadrature approximation (Equation (4)) into the original equation (Equation (3)), yields a set of ordinary differential equations to be solved by the usual numerical methods. The matrix of the weighting coefficients a_{ij} depends on the number of mesh points, N, and on the boundary conditions of the problem along the x-axis. For a given N and boundary conditions, the matrix a_{ij} can be uniquely determined using the various interpolating formulas with the criterion that the sum in Equation (1) should approximate Lu, 'best' in a certain sense. In this section we only consider the classical approach of using continuous polynomials, while in the later section we use the generalized approach of Sard (see Reference [8]) and spline polynomials.

Consider the nonlinear first-order partial differential equation

$$u_t = g(t, x, u, u_x), \quad -\infty < x < \infty, \quad t > 0$$

with initial condition

$$u(0, x) = h(x),$$

an equation arising in many mathematical models of physical processes. Let us make the assumption that the function is sufficiently smooth to allow us to write the approximate relation

$$u_x(t, x_i) \cong \sum_{j=1}^{N} a_{ij} u(t, x_j), \quad i = 1, 2, \cdots, N . \tag{5}$$

There are many ways of determining the coefficients a_{ij}. One method for determining these coefficients will be discussed below. Substitution of Equation (4) into Equation (3) yields the set of N ordinary differential equations

$$u_t(t, x_i) \cong g\left[t, x_i, u(t, x_i), \sum_{j=1}^{N} a_{ij} u(t, x_j)\right] \tag{6}$$

with initial conditions

$$u(0, x_i) = h(x_i), \quad i = 1, 2, \cdots, N . \tag{7}$$

Hence, under the assumption that (4) is valid, we have succeeded in reducing the task of solving Equation (3) to the task of solving a set of ordinary differential equations with prescribed initial values.

The numerical solution of such a system is a simple task for modern electronic computers using standard programs with minimal computing time and storage (in contrast to the case of partial differential equations). The additional storage needed for the solution of a partial differential equation using the differential quadrature technique is that of storing the N × N matrix a_{ij}. Considering that in practice it turns out that relatively low order differential quadrature is all that is needed, the total amount of storage and time required on the machine is quite low.

4. DETERMINATION OF WEIGHTING COEFFICIENTS

In order to determine appropriate coefficients in the approximation

$$u_x(x_i) \cong \sum_{j=1}^{N} a_{ij}u(x_j), \quad i = 1, 2, \cdots, N, \tag{8}$$

we can proceed by analogy with the classical quadrature case, demanding that Equation (8) be exact for all polynomials of degree less than or equal to N - 1. The test function $p_k(x) = x^{k-1}$, $k = 1, \cdots, N$, for arbitrary distinct x_i leads to the set of linear algebraic equations

$$\sum_{j=1}^{N} a_{ij}x_j^{k-1} = (k - 1)x_i^{k-2}, \quad k = 1, 2, \cdots, N, \tag{9}$$

which has a unique solution since the coefficient matrix is a Vandermonde matrix [5].

Rather than solving a set of linear algebraic equations, we may readily determine the a_{ij} explicitly once and for all if x_i are properly selected. In this discussion the x_i are chosen to be the roots of the shifted Legendre polynomial of degree N, $P_N^*(x)$ {with orthogonality range of [0, 1]}. The polynomials $P_N^*(x)$} are defined in terms of the usual Legendre polynomials by the relation

$$P_N^*(x) = P_N(1-2x). \tag{10}$$

By analogy with Lagrange's interpolation formula the test function is taken to be of the form

$$P_k(x) = \frac{P_N^*(x)}{(x-x_k)P_N^{*\prime}(x_k)}. \tag{11}$$

It follows that $p_k(x)$ is a polynomial of degree (N-1) such that $p_k(x_j) = \delta_{kj}$.

Using the fact that Equation (8) is to be exact for $u(x) = p_k(x)$, we see that

$$a_{ik} = \frac{P_N^{*\prime}(x_i)}{(x_i - x_k)P_N^{*\prime}(x_k)}; \; i \neq k \; . \tag{12}$$

For the case k=i, use of L'Hospital's rule plus the fact that $P_N^*(x)$ satisfies the differential equation

$$x(1-x^2)P_N^{*\prime\prime}(x) + (1-2x)P_N^{*\prime}(x)$$
$$+ N(N+1)P_N^*(x) = 0 \tag{13}$$

gives

$$a_{kk} = \frac{1-2x_k}{2x_k(x_k-1)} \; . \tag{14}$$

Using Equations (10), (12), and (14) the constant coefficients a_{ij} for any value of N, say N = 2, ..., 15, can be easily calculated. These values are then used as input data for a given problem once N has been chosen.

A number of partial differential equations were handled by this method in Reference [6] and numerical results obtained are in excellent agreement with those obtained by other techniques or by analytical means.

5. HIGHER ORDER PROBLEMS

We have seen that a good approximation to the first derivative of a function may often be obtained in the form

$$u_x(x_i) \cong \sum_{j=1}^{N} a_{ij}u(x_j), \; i=1, 2, \ldots, N \; . \tag{15}$$

Let us now indicate how this idea may be extended to higher order derivatives.

Viewing Equation (5) as a linear transforamtion of u,

$$u_x = Au \; , \tag{16}$$

we see that the second-order derivatives can be approximated by

$$u_{xx} = (u_x)_x = A(Au) = A^2u \,. \tag{17}$$

Thus higher order derivative approximations are obtained by iterating the linear transformation A.

When the method is applied to two and higher dimensional systems, the choice of N becomes critical. The previous results and the following numerical experiment show that low order approximations can be expected to yield both qualitative and quantitative information.

Let us now apply the foregoing ideas to the treatment of Burger's equation [7]

$$u_t + uu_x = \varepsilon u_{xx}, \; \varepsilon > 0 \tag{18}$$

with pure initial value condition

$$u(0, x) = f(x) \,. \tag{19}$$

Burger's equation enjoys a Riccati-like property in the sense that its solutions are expressible in terms of the solution of the linear heat equation

$$w_t = \varepsilon w_{xx} \,, \tag{20}$$

$$w(0, x) = g(x) \,, \tag{21}$$

by the transformation

$$u = -2\varepsilon W_x/W \tag{22}$$

$$f(x) = 2\varepsilon g_x(x)/g(x) \,. \tag{23}$$

This property will allow us to compare our numerical results with analytic solutions of Equations (20) to (23).

Adopting the notation

$$u_i(t) = u(t, x_i) \tag{24}$$

yields the set of ordinary differential equations

$$u_i'(t)+u_i(t)\sum_{j=1}^{N} a_{ij}u_j(t) = \varepsilon \sum_{k=1}^{N}\sum_{j=1}^{N} a_{ik}a_{kj}u_j(t) \quad i=1, 2, \ldots, N . \tag{25}$$

The initial conditions at t=0 are determined by the initial function

$$u_i(0) = f(x_i) \quad i=1, 2, \ldots, N . \tag{26}$$

The numerical experiments tried with given initial functions, when compared with the analytical results, proved the efficiency of the method adopted, see Reference [6].

6. SPLINE APPROXIMATION

In 1949, Sard [8] defined the best approximating function of order M, where $1 \leq M \leq N$, such that Equation (1) is exact for polynomial of degree (M-1) or less. About a decade later, Schoenberg [9] proved (among other properties) that the spline interpolating formula is the 'best' interpolation formula of order M in the sense of Sard [8].

Spline interpolating functions are a class of piecewise interpolating polynomial functions satisfying certain continuity properties at the interpolation points. Moreover, they are the 'smoothest' interpolating functions in the sense that the integral of Equation (27) is made as small as possible (see last section).

Let us now consider a subclass of spline called natural spline functions of degree (2M-1) which is defined (descriptive definition) by the following conditions:

(i) S is a polynomial of degree 2M-1 in each subinterval $[x_i, x_{i-1}]$, i=1, ..., N-1.

(ii) S is a polynomial of degree M-1 outside the region [a, b] (the boundary condition requirement).

(iii) $S, S', \cdots, S^{(2M-2)}$ are continuous at $x_1, \cdots, x_N$.

$$\text{(iv)} \quad S(x_i) = u(x_i); \quad i = 1, 2, \ldots, N \tag{27}$$

Schoenberg shows that the above function $S(x)$ is the unique solution of the minimum-interpolation problem as stated by Sard. In other words, an equivalent definition (the variational definition) of the natural spline is given as

(i) $\int_a^b [S^M(x)]^2\, dx$ exists and is minimized subject to the two following conditions

(ii) $S, S', \cdots, S^{(M-1)}$ are continuous in $[a, b]$.

(iii) $S(x_i) = u(x_i)$, $i=1, 2, \cdots, N$.

Both the above definitions of spline uniquely specifies the function $S(x)$ on the interval $[a, b]$. See references [9, 2]. As a result of the strong convergence properties of these functions, they can be very effectively used in approximate processes of interpolation, integration and differentiation.

Since for simplicity in our formulation we only consider the cubic spline, let us consider the case of $M = 2$ from now on. It is more convenient to formulate the problem in terms of a basis for the spline called cardinal (fundamental) spline. Cardinal spline is a spline for which exactly one defining value is one and all the others are zero. That is to say they are natural spline functions with the interpolating conditions

$$C_i(x_j) = \delta_{ij}, \qquad j=1, \ldots, N. \tag{28}$$

Then, the natural spline function S which interpolates $u(x)$ can be expressed as

$$S(x) = \sum_{i=1}^{N} u(x_i) C_i(x) \tag{29}$$

and the arbitrary function $u(x)$ can be approximated as

$$u(x) = \sum_{i=1}^{N} u(x_i) C_i(x) \,. \tag{30}$$

In the same manner, we obtain the best approximation of order M to a linear operator $Lu(x)$, Equation (1), by operating with L on the function $S(x)$ of Equation (29). That is to say

$$LS(x) = L \sum_{i=1}^{N} u(x_i)C_i(x) = \sum_{i=1}^{N} u(x_i)LC_i(x) \ . \qquad (31)$$

Comparison of the above equation with Equation (1) results in

$$a_i(x) = LC_i(x) \ . \qquad (32)$$

Therefore the weighting coefficients of the differential quadrature method can be uniquely determined from the above equation using natural cardinal splines, $C_i(x_j)$, $i, j=1, \cdots, N$, if the boundary conditions can be prescribed as $S''(x_1) = S''(x_N) = 0$ (called the natural boundary condition).

In general, the complete set of cardinal splines which spans the whole space depends on the boundary conditions of the problem. Let us consider the boundary condition of the partial differential equation to be given in terms of $u'(x_1) = \alpha_1$ and $u'(x_N) = \alpha_2$ and let us define a vector $\vec{y} = (u_1, u_1', u_2, \cdots, u_N, u_N')$. Then the N+2 cardinal splines which completely span the space are given as

$$\begin{aligned} &C(x_j) = \delta_{ij} && C_i'(x_1) = C_i'(x_N) = 0 \\ &C_0(x_j) = 0 && C_0'(x_j) = \delta_{1j} \\ &C_{N+1}(x_j) = 0 && C_{N+1}'(x_j) = \delta_{Nj} \end{aligned} \qquad (33)$$

for $i, j = 1, \cdots, N$. The spline interpolating function for this case is

$$S(x) = \sum_{i=1}^{N} u(x_i)C_i(x) + u_1'C_0(x) + u_N'C_{N+1}(x).$$

If we now consider the set of N+2 independent splines $C_k(x)$, $k=0, 1, \cdots, N+1$ and the vector y with elements y_k as defined above, then

$$S(x) = \sum_{k=0}^{N+1} y_k C_k(x) \tag{34}$$

and

$$LS(x) = \sum_{k=0}^{N+1} y_k [LC_k(x)] \tag{35}$$

so that the matrix of coefficients is obtained as

$$a_{ks} = [LS_k(x)]_{x=s} \quad k, s=0, 1, \cdots, M+1 , \tag{36}$$

For boundary conditions which are given in terms of the second derivatives, i.e., $u''(x_0) = \beta_1$ and $u''(x_M) = \beta_2$ the complete set of cardinal splines can be defined as

$$\begin{aligned} &C_i(x_j) = \delta_{ij} \qquad C_i''(x_1) = C_i''(x_M) = 0 \\ &C_0(x_j) = 0 \qquad C_0''(x_j) = \delta_{1j} \\ &C_{N+1}(x_j) = 0 \qquad C''_{N+1}(x_j) = \delta_{Nj} \\ &i, j=1, \cdots, N . \end{aligned} \tag{37}$$

Similarly, we can calculate the weighting coefficients for any general boundary conditions, using N+2 cardinal splines.

The cubic cardinal splines either with natural or general boundary conditions) can be uniquely determined from either of the two definitions of the spline functions described above. For the cubic case (M=2), the descriptive definition produces (4N-4) conditions which determine all of the 4(N-1) parameters of the cubic polynomials at all the sub-intervals $[x_i, x_{i+1}]$, i=1, ⋯, N-1, [2]. The evaluation of the cubic spline from the variational definition can be accomplished via dynamic programming [3]. This method is described in the last section in detail. Once the cardinal splines are computed for the general boundary conditions, the weighting coefficients are easily found and are stored for various values of N. Then any approximation on [a, b], for the use in differential quadrature, can be accomplished using the the stored values of a_{ij}. Usually, low values of N produce moderately accurate solutions, needed in many cases.

In the case of the periodic boundary conditions, periodic splines can be used. For a uniform mesh, calculations are simplified and only one cardinal spline must be computed with the rest being translates of this one; for details see Reference [2]. Many computations for solving partial differential equations, linear or nonlinear, can be carried out using these cardinal splines. Solutions of first or second order equations become quite simple since this method reduces the problem to the simple task of solving a set of algebraic equations.

7. APPROXIMATE SOLUTIONS

In trying to solve analytically the second order equations of the form

$$\varphi'' + [k^2 + v(x)]\varphi = 0 \tag{38}$$

we may be able in some cases to relate Equation (38) to some other well-known second order equations whose solutions are well-known. This can be achieved by suitable transformations of the dependent and the independent variable. In fact families of solvable potentials $v(x)$ for the Schrödinger equations were constructed in References [10], [11], and [12] with this idea. The general technique of generating solvable potentials is to take known eigenfunction polynomials and transform their differential equation to look like the Schrödinger equation in question. To illustrate the method, let us consider the equation

$$W''(z) + p(z)W' + VW = 0 \tag{39}$$

and make the transformation

$$Z = f(r); \; W(z) = g(r)\varphi(r) \tag{40}$$

and obtain the equation

$$\varphi''(r) + A(r)\varphi'(r) + B(r)\varphi(r) = 0 \, , \tag{41}$$

where

$$A(r) = \frac{d}{dr} \log(g^2/f'(r)) + p(f)f'(r) \tag{42}$$

$$B(r) = -\frac{d}{dr}\left[\frac{g'(r)}{g(r)}\right] - \left(\frac{g'(r)}{g(r)}\right)^2 + \frac{g'}{g}A(r) + q[f(r)]f'^2(r)\ . \tag{43}$$

Comparison of Equation (43) with the Schrödinger equation

$$\varphi'' + \left[k^2 - V(r) - \frac{\ell(\ell+1)}{r^2}\right]\varphi(v) = 0 \tag{44}$$

yields

$$A(r) = 0$$

$$B(r) = k^2 - V - \frac{\ell(\ell+1)}{r^2}\ . \tag{45}$$

A host of potentials v(r) can be generated by choosing the functions f and g compatible with (45). If we have started with Equation (39) whose solution is known, then the solution of the generated potential problem is achieved. The general philosophy of the method and the practical applications of this technique in the context of quantum mechanics is discussed in detail in Reference [12]. Of course, in the above procedure we are trying to get the exact solution with a given potential function. Suppose we are satisfied with approximate solutions of the equation

$$\varphi'' + f(x)\varphi = 0 \quad a \leq b$$
$$\varphi(a) = 0 \quad \varphi'(a) = 1 \tag{46}$$

Then we can approximate the function f(x) if we can, by another function g(x) in different regios in a piecewise fashion. If the solution of the equation

$$\varphi'' + g\varphi = 0 \qquad \varphi(a) = 0;\ \ \varphi'(a) = 1 \tag{47}$$

is known, then the approximate solution of Equation (46) is given by the known analytic solution fo (47) in the different regions. For example, let the function f(x) be such that we could approximate it by straight lines in five segmental intervals, aa_1, a_1a_2, a_2a_3, a_3a_4, a_4a_5, a_5b. This means that in the successive intervals f(x) is to be replaced by $g_1(x) = \ell_1+m_1x$; $g_2(x) = \ell_2+m_2x\cdots$ etc. The solution in the first interval is the solution of the equation

$$\bar{\varphi}_1''(x) + (\ell_1 + m_1 x)\bar{\varphi}_1(x) = 0; \; \bar{\varphi}_1(a) = 0, \quad \varphi'(a) = 1 \tag{48}$$

This is known analytically and hence for the second interval the Equation (48) is substituted by

$$\bar{\varphi}_2'' + (\ell_2 + m_2 x)\bar{\varphi}_2(x) = 0; \; \bar{\varphi}_2(a_1) = \bar{\varphi}_1(a_1) \; \bar{\varphi}_2'(a_1) = \bar{\varphi}_1'(a_1) \tag{49}$$

and so on.

The initial values for the successive segments are obtained from the solution valid in previous segments. Thus we have the solution of Equation (46) over the entire interval approximated by the known solutions of (47) in the various segments. The closeness of this solution to the actual solution of (46) depends on how close the $g(x)$ in various segments approximates the original function $f(x)$.

The segmental approximation of $f(x)$ by fitting straight lines, or a polynomial of higher order, can be accomplished by dynamic programming techniques. The main idea in fitting any desired function to the curve in a piecewise fashion is to make the mean square norm of the deviation of the curve from the given function in each section as small as possible. It may also be the maximum norm of the deviation. In the usual spline approximation it is the curvature norm. In all the cases the minimization holds locally in the segment as well as over the entire region. Hence, the optimal choosing of the coefficients in each interval can be done in a sequential fashion using the methods of dynamic programming. Fitting the curve with line segments carried out by Bellman and Roth [14] and mean square spline approximation (see Reference [15]) methods are given in the next section.

Depending on the nature of the function $f(x)$, the approximating function $g(x)$ has to be chosen. It should be such that we should have an analytic solution of Equation (47). For example, if we have straight line approximation in each segment $g(x) = \ell + mx$, and Equation (47) becomes

$$\varphi''(x) + (\ell + mx)\varphi(x) = 0$$

$$\varphi(a) = 0; \; \varphi'(a) = 1 \, . \tag{50}$$

Let us make the transformation

$$(x + \ell/m)m^{1/3} = \xi \, . \tag{51}$$

Then Equation (50) becomes

$$\frac{d^2W}{d\xi^2} + \xi W = 0 \; . \tag{52}$$

The two independent solutions of Equation (52) are the Airy functions Ai(-ξ) and Bi(-ξ). These are linear combinations of Bessel functions of order 1/3, i.e., $J_{\pm 1/3}$.

$$Ai(-z) = \frac{1}{3}z^{1/2}\left[J_{1/3}\left(\frac{2}{3}z^{3/2}\right) + J_{-1/3}\left(\frac{2}{3}z^{3/2}\right)\right]$$

$$Bi(-Z) = \frac{1}{3}Z^{1/2}\left[J_{-1/3}\left(\frac{2}{3}Z^{3/2}\right) - J_{1/3}\left(\frac{2}{3}Z^{3/2}\right)\right] . \tag{53}$$

Thus the approximate solutions of the equation in each interval are known.

Similarly if g(x) is given by a second degree polynomial such as

$$g(x) = \ell + m_1 x + m_2 x^2 \tag{54}$$

the equation

$$W'' + (\ell + m_1 x + m_2 x^2)W = 0 \tag{55}$$

can be related by a series of transformations to a differential equation whose solution involves the Kummer from of the confluent hypergeometric function.

If g(x) happens to be of the form

$$g(x) = \frac{1}{(\ell x + m)^2} \; , \tag{56}$$

we know already from our discussions of Chapter V that

$$W'' + g(x)W = 0 \tag{57}$$

has exact solutions which are given by the WKB solutions corresponding to (57).

$$W_{\pm} = \frac{1}{[g(x)]^{1/4}} \exp\left(\pm i\int_a^x \sqrt{g(s)}\, ds\right). \tag{58}$$

Thus we can multiply the examples of various approximation functions which essentially have to be determined by the nature of the actual function f(x).

8. SEGMENTAL CURVE FITTING

As pointed out in earlier sections, if the function f(t) can be approximated by different curves in different regions it might be a better approximation than fitting a single polynomial over the entire region. We have explained in the first section the spline approxiamtion which corresponds to piecewise curve fitting. A cruder approximation might be to fit different straight lines in different regions.

Let us now see how best we could fit a given set of data corresponding to a function v(t) in a piecewise fashion. This is a simpler problem than even constructing a cubic spline in each region.

Bellman and Roth [14] consider such a problem using the criterion that the best fit is that which minimizes the maximum absolute deviation between the fitted curve and the given data. A least square criterion might also have been used. In any case, if a function g(t) is given with its values on the interval [0, a], we wish to find the most suitable straight line segments, each connecting its neighbor at its end points satisfying the criteria regarding the maximum deviation from the actual curve. The problem is to determine the end points of each segment in such a way that the final curve best approximates g(t) over the entire interval [0, a]. We can apply the method of dynamic programming to decide on the choice of the end points of the segments in a sequential fasion.

If any two points in the domain of definition of g(t) have the coordinates (t_i, S_{ij}) and (t_k, S_{kl}), then a straight line conneting these points can be written as

$$y(t) = \frac{(S_{ij} - S_{kl})}{(t_i - t_k)} (t - t_k) + S_{kl} . \tag{59}$$

Let the measure of the error between the given curve g(t) and its straight line approximation y(t) in the interval $[t_i, t_k]$ be simply defined as

$$D(S_{ij}, S_{kl}, t_i, t_k) = \max_{t_k \le t \le t_i} \left| g(t) - \left\{ \frac{(s_{ij} - S_{kl})}{(t_i - t_k)} (t-t_k)+S_{kl} \right\} \right| . \tag{60}$$

The problem is to determine the end points which, when connected, give the minimum error over (0, a) between g(t) and the fitted curve.

9. DYNAMIC PROGRAMMING APPROACH

In the spirit of dynamic programming, let the functional $F_N(b, a)$ be the minimum error in fitting the curve g(t) over the interval [0, a] by N segmented striaght lines when the right end point of the last straight line segment is located at the point (a, b). The point is that we wish to replace this single stage problem by a multistage process in which first the last joint point on the right is chosen, then the next to last, and so on. Then for M > 1, the principle of optimality yields the functional equations

$$F_M(S_{ij}\ t_i) = \min_{S_{ki};t_1<t_k<t_i} \max\ [D(S_{ij}S_{kl}t_it_k) + F_{M-1}(S_{kl}t_k)]$$

$M = 2, \ldots, N.$

$$F_1(S_{ij}t_i) = \min_{S_{1\ell}} [D(S_{ij}S_{1\ell}t_it_1)]. \tag{61}$$

The computational problem is to determine $F_N(b, a)$ and the solution is determined by that value of b for which $F_N(b, a)$ is a minimum. A byproduct of this solution will be the set of end points which will determine the best sequence of striaght line segments approximating the curve g(t) over the interval [0, a].

As in the standard dynamic programming practice, we will first evaluate the functionals in the order $F_1, F_2, \ldots, F_N$, and will then proceed to extract the solution in the reverse order. The computational procedure is given in Reference [13].

In a similar fashion, if we are to fit a quadratic polynomial using the criterion that the square of the derivative is a minimum in each segment, we can take the functional

$$\sum_{j=1}^{N+1} \int_{u_{j-1}}^{u_j} (s(x) - h(x, a_jb_jc_j)^2\ dx \tag{62}$$

where $h(x, a_jb_jc_j) = a_j + b_jx + c_jx^2$.

The coefficients a_j, b_j, c_j are obtained by a minimization procedure using the dynamic programming method explained in the last section for a cubic polynomial using a different criterion. u_j's are the segmental points. We refer the reader to the paper on mean square splines by Bellman et al. [15].

10. SPLINES VIA DYNAMIC PROGRAMMING

In this section we derive the spline approximation using the dynamic programming technique. The spline approximate a function in a piecewise fashion by means of a different polynomial in each subinterval. The cubic spline has its origins in beam theory. It possesses continuous first and second derivatives at the knots and is characterized by a minimum curvature property [2] which is related to the physical feature of minimum potential energy of the supported beam.

Translated into mathematical terms, this means that between successive knots the approximation yields a third order polynomial with its first two derivatives continuous at the knot points. The minimum curvature property holds good for each subinterval as well as for the whole region of approximation. This means that the integral of the square of the second derivative over the entire interval, and also over each subinterval, is to be minimized. Thus, the task of determining the spline offers itself as a textbook problem in discrete dynamic programming [19]. Starting with the initial value of the function and assuming an initial slope of the curve, the minimum norm property of the curvature makes sequential decision of the slope at successive knot points feasible.

It is the aim of this chapter to derive the cubic spline by the methods of dynamic programming and show that the results obtained lead to recursion relations which are equivalent to the usual relations obtained for the so-called 'moments' in successive intervals of the spline approximation [2], [3].

11. DERIVATION OF SPLINE BY DYNAMIC PROGRAMMING

In deriving the spline approximation, to a function f(x) is an interval $a \leq x \leq b$, we subdivide it into a number of mesh points corresponding to the knots

$$\Delta : a = x_0 < x_1 < x_2 \ldots < x_n = b \,. \tag{63}$$

The set of values the function $f(x)$ takes is prescribed

$$y : y_0 y_1, \ldots, y_n . \tag{64}$$

We seek an approximation to $f(x)$ denoted by $S_{\Delta,y}(x)$ which is continuous with its first and second derivatives on [a, b], coincides with a cubic in each interval $x_{j-1} \le x \le x_j$ $(j=1, 2, \ldots, n)$ and satisfies the condition

$$S_{\Delta,y}(x_j) = y_j \quad j=0, 1, 2, \ldots, n. \tag{65}$$

Holladay's theorem (see Reference 16) asserts that the the spline function $S_{\Delta,y}(x)$ with $S''_{\Delta,y}(a) = S''_{\Delta,y}(b) = 0$ minimizes the integral

$$J(a, b) = \int_a^b [(f''(x)]^2 \, dx . \tag{66}$$

Having thus spelled out the boundary conditions and the interpolating values of the $f(x)$ at the knots as well as the criterion (66) characterizing the approximation, we will apply the methods of dynamic programming to determine $S(x)$ in each subinterval. To this end let us consider the interval $x_i \le x \le x_{i+1}$ and write $S_i(x)$ as

$$S_i(x) = y_i + m_i(x-x_i) + p_i(x-x_i)^2 + q_i(x-x_i)^3 \quad \text{for} \quad x_i \le x \le x_{i+1}. \tag{67}$$

Since y_i and y_{i+1} are known, letting m_i and m_{i+1} denote the first 'moments' at the knots x_i and x_{i+1} we find, using the continuity properties, that

$$P_i = \frac{1}{h_{i+1}} [3r_i - (m_{i+1} + 2m_i)] \tag{68}$$

$$q_i = \frac{1}{h_{i+1}^2} [(m_{i+1} + m_i) - 2r_i] , \tag{69}$$

where

$$r_i = (y_{i+1} - y_1)/h_{i+1} \tag{70}$$

and

$$h_{i+1} = x_{i+1} - x_i \ . \tag{71}$$

IN order to arrive at the multistage decision process for the determination of the moments at successive knots, we write the minimum value of the integral corresponding to the minimum curvature property, starting with the knot point x_i as:

$$F_i(m_i) = \underset{\substack{m_{i+k} \\ k=1,\ \ldots,\ n-i}}{\text{Min}} \sum_{j=i}^{n-1} \int_{x_j}^{x_{j+1}} [S_i''(x)]^2 \, dx \ ; \ x_n = b. \tag{72}$$

Here we have assumed that the first moment at x_i is m_i. All the subsequent values of the slopes $m_{i+k} (k \geq 1)$ are to be found in terms of m_i by the minimization condition. Equation (72) in the usual fasion, see [3], leads to the recurrence relation

$$\begin{aligned} F_i(m_i) &= \underset{m_{i+1}}{\text{Min}} \left[\int_{x_i}^{x_{i+1}} [S_i''(x)]^2 \, dx + F_{i+1}(m_{i+1}) \right] \\ &= \underset{m_{i+1}}{\text{Min}} \left[g_{i+1}\{3r_i^2 + m_{i+1}^2 + m_i^2 - 3r_i(m_{i+1}+m_i) + \right. \\ &\qquad \left. + m_i m_{i+1}\} + F_{i+1}(m_{i+1}) \right]. \end{aligned} \tag{73}$$

with

$$g_{i+1} = 4/h_{i+1} \ . \tag{74}$$

It is easy now to argue that $F_i(m_i)$ is quadratic in m_i. To see this let us take the last interval $x_{n-1} \leq x \leq x_n$ where

$$F_{n-1}(m_{n-1}) = g_n \left[\frac{3}{4} r_{n-1}^2 - \frac{3}{2} r_{n-1} m_{n-1} + \frac{3}{4} m_{n-1}^2 \right] . \tag{75}$$

The above expression is clearly quadratic in m_{n-1}. Hence by induction, or otherwise, it can be shown that F_i is quadratic in m_i. Therefore we may write

$$F_i(m_i) = c_i + d_i m_i + e_i m_i^2 \,. \tag{76}$$

From Equation (73) we obtain

$$F_i(m_i) = \underset{m_{i+1}}{\text{Min}} \Big[[3r_i^2 + m_{i+1}^2 + m_i^2 - 3r_i(m_{i+1} + m_i) + m_i m_{i+1}] g_{i+1} + c_{i+1} + d_{i+1} m_{i+1} + e_{i+1} m_{i+1}^2 \Big] \,. \tag{77}$$

The minimization over m_{i+1} is readily performed. Thus Equations (76) and (77) lead to the following recursion relations

$$m_{i+1} = \frac{3r_i g_{i+1} - g_{i+1} m_i - d_{i+1}}{2(g_{i+1} + e_{i+1})} \tag{78}$$

$$c_i = c_{i+1} + 3r_i^2 g_{i+1} - \frac{(3r_i g_{i+1} - d_{i+1})^2}{4(g_{i+1} + e_{i+1})} \,, \tag{79}$$

$$e_i = g_{i+1} \left[1 - \frac{g_{i+1}}{4(g_{i+1} + e_{i+1})} \right] , \tag{80}$$

$$d_i = g_{i+1} \left[\frac{3r_i g_{i+1} - d_{i+1}}{2(g_{i+1} + e_{i+1})} - 3r_i \right] . \tag{81}$$

From Equation (75) the values of c_{n-1}, e_{n-1}, and d_{n-1} are given by

$$\begin{aligned} c_{n-1} &= 3r_{n-1}^2/h_n \\ d_{n-1} &= -6r_{n-1}/h_n \\ e_{n-1} &= 3/h_n \,. \end{aligned} \tag{82}$$

Hence, c_i, d_i, and e_i for $i = 0, 1, \ldots, n-1$ can be easily obtained from the recursive relations. Similarly, all the m_i, i=0, 1, ..., n can be obtained in terms of the assumed value m_0 of the slope at x = a. The minimum value assumes the form

$$F_0(m_0) = c_0 + d_0 m_0 + e_0 m_0^2 \ . \tag{83}$$

The value of m_0, which minimizes this expression, is the starting value of the recursion equation for the m_i. Once these quantities are known, use of Equations (68) and (69) will yield the spline functions S_i in each interval. The relations are simple and the calculations can be performed on the computer in a very short time.

12. EQUIVALENCE OF THE RECURSION RELATIONS OBTAINED BY DYNAMIC PROGRAMMING AND THE USUAL RESULTS

Let us consider the recursive relations for m_i and the coefficients c_i, e_i and d_i earlier. We can easily rewrite the recursive relations as

$$m_{i+1} = \frac{d_i}{g_{i+1}} + 3r_i - \frac{g_{i+1}m_i}{2(g_{i+1} + e_{i+1})} \ . \tag{84}$$

Also

$$2(g_i + e_i)m_i = 3r_{i-1}g_i - d_i - g_i m_{i-1} \ . \tag{85}$$

Adding these two relations, and using Equation (80) we have

$$\lambda_i m_{i-1} + 2m_i + \mu_i m_{i+1} = 3\lambda_i \frac{[y_i - y_{i-1}]}{h_i} + 3\mu_i \frac{[y_{i+1} - y_i]}{h_{i+1}} \tag{86}$$

where $\lambda_i = h_{i+1}/(h_i + h_{i+1})$ and $\mu_i = 1 - \lambda_i$.

This is precisely the recursive relation obtained by the usual method (Reference [2]).

The end relations in our case are given by the minimization procedure we adopt. Thus at x = b, minimizing $F_{n-1}(y_{n-1})$ yields the relation

$$2m_n + m_{n-1} = \frac{3[y_n - y_{n-1}]}{h_n} . \tag{87}$$

At the other end x = a, minimizing $F_0(m_0)$ gives

$$m_0 = - \frac{d_0}{2e_0} . \tag{88}$$

We know that

$$m_1 = \frac{3r_0 g_1 - g_1 m_0 - d_1}{2(g_1 + e_1)} \tag{89}$$

where $r_0 = (y_1 - y_0)/h_1$ and $g_1 = 4/h_1$ with $h_1 = (x_1 - x_0)$.

Making use of the relations obtained earlier for d and e, and suing Equations (88) and (89), we have

$$m_1 + 2m_0 = 3r_0 \quad \text{at } x = a . \tag{90}$$

These two end conditions express the fact that p_0 and p_n, the second moments at the end points, are equal to zero.

Thus, we have found that our recursive relations are equivalent to the usual results [2], corresponding to the end conditions mentioned earlier. Any other end conditions can be taken care of by the dynamic programming procedure by altering the criterion function in a suitable way.

The spline approximation for functions of two variables called the bicubic spline can be obtained by the dynamic programming methods also. The reader can look up References [17], [18] with profit. With this spline approximation we are able to solve partial differential equations occurring in various fields and furthermore we can make use of this technique in system identitification problems in a very effective manner [20].

These methods can be extended to derive the cardinal splines, polynomial splines of odd orders, 5, 7, etc., by employing the dynamic programming methods. 'Generalized splines' also can be obtained and in each case we are led to the recursive set of equations for the coefficients of the approximation function in each subinterval. Computation of these offer no problems since the recursive set of equations form a stable system.

REFERENCES

[1] Bellman, R. and J. Casti, 'Differential Quadrature and Long Term Integration', Journal of Mathematical Analysis and Applications 34 (1971), 235.
[2] Ahlberg, J.H., E.N. Nilson, and J.L. Walsh, The Theory of Splines and Their Applications, Academic Press, Inc., 1967.
[3] Bellman, R., B.G. Kashef, and R. Vasudevan, 'Splines via Dynamic Programming', Journal of Mathematical Analysis and Applications 38 (1972), 47.
[4] Bellman, R., R. Kalaba, and J. Lockett, Numerical Inversion of the Laplace Transform, American Elsevier Publishing Co., Inc., New York, 1966.
[5] Bellman, R., H. Kagiwada, R. Kalaba, and M. Prestrud, Invariant Imbedding and Time Dependent Transport Processes, American Elsevier Publishing Co., Inc., New York, 1964.
[6] Bellman, R., B.G. Kashef, and J. Casti, 'Differential Quadrature: A Technique for Rapid Solution of Nonlinear Differential Equations', Journal of Computational Physics 10 (1972).
[7] Hopf, E., Comm. in Pure and Applied Mathematics 3 (1950), 201.
[8] Sard, A., 'Linear Approximation', Math. Survey 9, American Math. Society Publ., Rhode Island (1963).
[9] Schoenberg, I.J., 'On the Best Approximation of Linear Operators', Indag. Math. 26 (1964), 155.
[10] Bargman, V., Rev. Mod. Phys.21 (1949), 488.
[11] Morse, P.M. and H. Feshbach, Methods of Theoretical Physics: Part II, McGraw-Hill, New York, 1953.
[12] Battacharjee, A.C. and E.C.G. Sudarshan, Nuovo Cimento 25 (1962), 864.
[13] Vasudevan, R., K. Venkatesan, and G. Jagannathan, 'Construction of Solvable Potentials and Some Aspects of Regularization of Singular Potentials', Supplimento al Nuovo Cimento 5 (1967), 621.
[14] (a) Bellman, R., 'On the Approximation of Curves by (line) Segments Using Dynamic Programming', Communications of ACM 4 (1961), 284.
(b) Bellman, R. and R. Roth, 'Curve Fitting by Segmented Straight Lines', Journal American Stat. Assoc. 64 (1969), 3.
[15] Bellman, R., B. Kashef and R. Vasudevan, 'Mean Square Spline Approximation', Journal of Mathematical Analysis and Applications 45 (1974), 47-53.

[16] Holladay, J.C., 'Smoothest Curve Approximation', Math. Tab. Aids Comm. 11 (1957), 233.
[17] De Boor, C., 'Bicubic Spline Approximation', Journal of Mathematics and Physics 4 (1962), 41.
[18] Vasudevan, R., R. Bellman, and B.G. Kashef, 'Dynamic Programming and Bicubic Spline Interpolation', Journal of Mathematical Analysis and Applications 44 (1973), 160-174.
[19] Bellman, R., Dynamic Programming, Princeton University Press, Princeton, N.J., 1957.
[20] (a) Bellman, R., B. Kashef, and R. Vasudevan, 'Partial Differential Equations, Splines and Differential Quadrature', University of Southern California Preprint.
(b) 'The Inverse Problem of Estimating Heart, Parameters from Cardiograms', Math. Biosciences 19 (1974), 221.
[21] Bellman, R., Stability Theory of Differential Equations, Dover Publications, New York, 1969.
[22] Bellman, R., Method of Nonlinear Analysis, Vol. I and Vol. II, Academic Press, Inc., New York, 1970, 1973.

EXERCISES AND NOTES

CHAPTER I

1. Show that if $u_1(t)$ is a nontrivial solution of $u''(t) - a(t)u(t) = 0$

$$u_2(t) = u_1(t) \int_{t_0}^{t} [u_1(s)]^{-2}\, ds$$

is another linearly independent solution.

2. Given the scalar differential euqation

$$u''(t) + \frac{4}{t} u' + \frac{2}{t^2} u = \frac{\sin t}{t}, \quad t > 0$$

show that $u_1(t) = \frac{1}{t}$; $u_2(t) = \frac{1}{t^2}$ constitute a fundamental system of solutions of the corresponding homogeneous equation. Then using the variation of constants formula find the solution $u(t)$ of the given equation satisfying the initial conditions $u(1) = 1$, $u'(1) = 0$.

3. Given that $f(t)$ is continuous on $r_1 < t < r_2$, the solution of $u^{(n)} = f(t)$ is given by

$$u(t) = \int_0^t f(s) \frac{(t-s)^{n-1}}{(n-1)!}\, ds$$

for the initial conditions given as $u(0) = u'(0) = \cdots = u^{n-1}(0) = 0$.

4. Consider a potential barrier for a quantum mechanical particle in one dimension. Let us say the potential V(x) is given by

$$V(x) = 0, \quad x < -a$$

$$= V_0, \quad -a < x < a$$

$$= 0, \quad a < x$$

If the energy of the particle $E < V_0$, no penetration of the barrier will occur in classical physics. However, in quantum mechanics we can find some amount of transmission probability of the wave, even though the energy of the particle lies below the top of the barrier. Find out the transmission probability as

$$|T|^2 = \frac{(2kK)^2}{(k^2 + K^2)^2 \sin h^2 2Ka + (2kK)^2},$$

where $k^2 = E$, and $i\,K^2 = i\,(2m/\hbar^2)\,(V_0 - E)$.

5. Show that for the wave equations

$$\Psi'' + Q^2(x)\Psi = 0$$

the WKB solution is the exact solution when Q^2 is given by the expression

$$Q^2 = (Ax + B)^{-4},$$

where A and B are arbitrary constants.

CHAPTER II

1. If u_1 and u_2 are two functions with requisite number of derivatives, show that

$$\begin{vmatrix} u & u' & u'' \\ u_1 & u_1' & u_1'' \\ u_2 & u_2' & u_2'' \end{vmatrix} = 0$$

is a linear differential equation with two particular solutions u_1 and u_2 .

2. Show that

$$\frac{d^2u}{dt^2} + a_1(t)\frac{du}{dt} + a_2(t)u = \frac{w}{u_1}\frac{d}{dt}\left(\frac{u_1^2}{w}\frac{d}{dt}\left(\frac{u}{u_1}\right)\right),$$

where u_1 and u_2 are two linearly independent solutions of the homogeneous equation and w is the Wronskian of u_1 and u_2 (for derivation of this result and extensions see Chapter IV of Quasilinearization and Nonlinear Boundary Value Problems, by R. Bellman and R. Kalaba, American Elsevier Publ. Co., New York, 1965).

3. Obtain the solution of the equation

$$\frac{d^2u}{dt^2} + a(t)\frac{du}{dt} + b(t)u = v; \quad u(0) = c_1 \quad u'(0) = c_2$$

by variation of parameters as

$$u = c_1u_1 + c_2u_2 + \int_0^t v(s)\exp\left(\int_0^s a(s_1)ds_1\right)[u_1(s)u_2(t) - u_2(s)u_1(t)]\,ds,$$

where u_1 and u_2 are the solutions of the homogeneous equation with

$$u_1(0) = 1; \quad u_1'(0) = 0; \quad u_2(0) = 0; \quad u_2'(0) = 1.$$

4. The idealized netron-proton interaction is of the form $v(r) = -Ae^{-r/a}$. The reduced Schrödinger equation for this one-body problem may be written as

$$\frac{d^2x}{dr^2} + \frac{2m}{\hbar^2}(E + Ae^{-r/a})x = 0$$

with the boundary conditions, $x(\infty) = 0$ and $x(0) \to 0$. Use the variable $y = e^{-r/29}$ in the place of r. Find the appropriate solution.

5. Determine the asymptotic behavior of $u'' + u/t = 0$ by applying Liouville transformations, repeatedly.

6. Consider the equation $u'' + q(t)u = 0$ and make the simultaneous change of dependent and independent variables $v = u'$; $ds = q(t)\,dt$. Show that

$$\frac{d^2v}{ds^2} + \frac{v}{q(t)} = 0$$

and thus consider the connection between the behavior of the solutions of $u'' + a^2u = 0$ as $a(t) \to 0$ and as $a(t) \to \infty$ (see P. Hartman, 'On the Existence of Large and Small Solutions of Linear Differential Equations', Technical Note number 20, John Hopkins University Press, 1960, and A. Wintner, 'On a Principle of Reciprocity Between High-Frequency and Low-Frequency Problems', Quarterly of Applied Mathematics 15 (1957), 314.

7. For the second order differential equation

$$\frac{d^2u}{dz^2} + Q^2(z)u = 0$$

by the following transformations,

$$u(z) = q^{-1/2}\varphi; \quad \text{and} \quad w(z) = \int q(\xi)\,d\xi$$

if it is possible that one can arrive at

$$\frac{d^2\varphi}{dw^2} + \varphi = 0, \text{ and } u = [q(z)]^{-1/2} \exp\left[\pm i \int^z q(\xi)\,d\xi\right]$$

(a) show that q satisfies the equation

(1) $q^{-1/2}\dfrac{d^{-2}q^{-1/2}}{dz^2} = q^2 - Q^2$ and

(2) $iq' = 2q(Y-q)$,

where iY is the logarithmic derivative of the solution,

(b) as shown in this chapter if we start with the equation

$$\frac{d^2u}{dz^2} + \lambda^2Q^2 = 0 \quad \text{where } \lambda \text{ is a large parameter}$$

and if $u = \exp\left[i \int^z Y dl\right]$

and if Y is expanded in powers of λ^{-1} as

$$Y(z) = \sum_{n=-1}^{\infty} Y_n \lambda^{-n} \quad \text{with } Y_{-1} = \pm Q$$

show that

(1) $iY_n' = \sum_{\nu=-1}^{n+1} Y_\nu Y_{n-\nu}$, $n=0, 1, \ldots$ and that

(2) $q(z) = \sum_{k=0}^{\infty} Y_{2k-1} \lambda^{-2k+1}$.

This can be shown by splitting of $Y = Y_{odd} + Y_{even}$

(c) Also it can be shown

$$iY'_{even} = Y^2_{even} + Y^2_{odd} - \lambda^2 Q^2$$

and for the odd part we have

$$iY'_{odd} = 2Y_{odd}\, Y_{even} .$$

See, for details, the paper entitled 'Note on WKB Method' by R.K. Nesbet, Journal of Mathematical Physics 9 (1968), 205.

CHAPTER III

1. In the text the imbedding has been done at the end of the rod where the beam is incident. Do the imbedding at the opposite end and obtain the equations for the reflection and transmission functions by particle counting for the single state model.

2. Suppose the coefficients in $W' = a(Z) + b(Z)W + C(Z)W^2$ are analytic functions of Z in some circle $|Z|<r$. Write the solution of the problem as $W(Z) = a_0 + a_k Z^k/S(Z)$. Find the Riccati equation satisfied by S(Z). Hence, show that the solution W can be obtained as a continued fraction

$$W(Z) = a_0 + \cfrac{a_k Z^k}{b_0 + \cfrac{b_m Z^m}{c_0 + \cdots}} .$$

3. Consider the second order equation $y'' = y + 1$ with $y(0) = \alpha$ and $y'(x) = \beta$. Take $u = y$ and $y' = v$ and write the coupled transport equation and obtain the solution using R and T functions.

4. Given the problem $y'' + a(t)y = 0$. $y(0) = c_1$. $y'(x) = c_2$. This can be reduced to the imbedding equations we have been considering. Consider the following iterative scheme. Let $y_0(0) = c_1$ and $y_0'(0) = b_0$. b_0 arbitrary, calculate $y_0(x) = b_1$. Now set $y_1(x) = b_1$ and $y_1'(x) = c_2$ compute $y_1'(0) = b_2$. Set $y_2(0) = c_1$ and $y_2'(0) = b_2$ and so on. Does this procedure converge? (See R. Bellman and T. Brown, 'On the Iterative Solution of Two Point Boundary Value Problems', Boll. D'unione Matimatico 16 (1961), 285-287.)

5. Consider the transport equations for u and v in a medium with a reflecting medium attached at the end $Z = 0$. The value of u at $Z = 0$ is given in terms of the value of v at that end, i.e., $u(0) = \beta v(0)$. The reflector returns on the average β particles for each particle emerging at the left end. The condition on the right end remains the same $v(x) = 1$. Compute the flux emerging from right end by imbedding methods.

6. Consider the transport of particles in the multistate model, in a medium with absorption properties. Arrive at the matrix differential equations for R and T, the usual reflection and transmission matrices. Let $L(x)$ be the matrix representing the total absorption of the medium of particles in different states. Assume $e_{ii}\Delta$ to be the probability that the particle in the i-th state gets absorbed in Δ while moving to the right and $f_{ii}\Delta$ while moving to the left. Introduce functions $\ell_{ij}(a)$ = the probability that a particle of j-th type incident from the left results in the absorption of a particle of type i in the medium $[a, 0]$. $i, j = 1, 2, \ldots, N$. $E(e_{ii})$ and $F(f_{ii})$ are diagonal matrices. Obtain the differential equation satisfied by $L(x)$. What does the conservation conditions mean in this case? (See R. Bellman, 'Scattering Processes and Invariant Imbedding', Journal of Mathematical Analysis and Applications 23 (1968), 254.)

7. Show that if an obstacle is represented by

$$Q = \frac{1+d}{2T_r} - \frac{R_r+R_\ell}{2T_r}\sigma_1 + \frac{R_r-R_\ell}{2iT_r}\sigma_2 - \frac{1-d}{2iT_r}\sigma_3 ,$$

where $d = T_r T_\ell - R_r R_\ell$ and σ_1, σ_2 and σ_3 are the elements of a quarternion algebra, the star product rule for $(T_r, R_r, T_\ell, R_\ell)$ makes these symbols combine by the rules of quarternion multiplication.

8. If we consider two networks that permute, i.e., that they give the same result independently of their order in the line, the star product rules yield the result that

$$\frac{T_r T_\ell - R_r R_\ell - 1}{R_\ell} \text{ and } \frac{R_r}{R_\ell}$$

should be the same for each one of constituent networks.

CHAPTER IV

1. Consider a multilayer medium with layer interfaces at $x_0 < x_1 < x_2 < x_3 \cdots < x_N < x_{N+1}$. The space to the left of x_0 has a wave number k_0 and that to the right of x_{N+1} has a number of k_{N+1} and medium in the stratum $x_{j-1} \leq z < x_j$ is characterized by the wave number k_j. Using the imbedding approach, show that ρ_{N+1} and τ_{N+1} the reflection and transmission amplitudes respectively for unit incident wave at $z = N+1$ can be related to ρ_N and τ_N recursively

$$\rho_{N+1} = \frac{k_{N+2}-k_{N+1}}{k_{N+2}+k_{N+1}} + \frac{4k_{N+2}k_{N+1}}{(k_{N+2}+k_{N+1})^2} \frac{e^{(2ik_{N+1}\Delta_{N+1})}\rho_N}{1+\left(\frac{k_{N+2}-k_{N+1}}{k_{N+2}+k_{N+1}}\right)e^{(2ik_N\Delta_{N+1})}\rho_N},$$

also

$$\tau_{N+1} = \frac{2k_{N+2}}{k_{N+2}k_{N+1}} \frac{\exp(ik_{N+1}\Delta_{N+1})\tau_N}{1+\exp(2ik_{N+1}\Delta_{N+1})\left(\frac{k_{N+2}-k_{N+1}}{k_{N+2}+k_{N+1}}\right)\rho_N}.$$

This result was retained by Stokes (Mathematical and Physical Papers 2 Cambridge Univ. Press, 1983).

2. Consider a wave traveling through a medium with wave number $k(z)$. Suppose in passing through a stratum Z and Z+Δ

there is a probability $p(z)\Delta$ that the wave will be absorbed but in return will produce two waves of the same amplitude and phase one traveling to the left and the other traveling to the right. The probability that such a fission does not occur in Δ is given by $(1 - p(z)\Delta)$. Show that the reflection R for a right moving wave impinging on the medium extending from z to b is

$$\frac{dR}{dz} = \frac{k'(z)}{2k(z)} - p(z) - 2ik(z)R(z) - \left(\frac{k'}{2k} + p\right)R^2(z)$$

what is the initial condition on R.

3. Using quasilinearization techniques obtain the first few approximate solutions for $\rho(x)$ from Equation (32). Compare them with other known appriximations such as WKB and Born approximations.

4. If all the solutions of $dy/dt = Ay$ where A is a constant matrix are bounded as $t \to \infty$ the same is true of the solutions of $dz/dt = [A + B(t)]z$, provided $\int_0^\infty \|B(t)\| dt < \infty$. (Use Bellman-Gronwell inequality.)

5. If (a) $\varphi(t) \to 0$ as $t \to \infty$ and (b) $\int_0^\infty \varphi^2(t)\, dt < \infty$ there are two solutions of $u'' - (1 + \varphi(t))u = 0$, possessing the asymptotic forms

$$u_1 = \exp\left\{t + \frac{1}{2}\int_0^t \varphi(t)\, dt + O(1)\right\}$$

$$u_2 = \exp\left\{-t - \frac{1}{2}\int_0^t \varphi(t)\, dt + O(1)\right\}.$$

CHAPTER V

1. Derive equation for ρ and τ, the reflection and transmission coefficients at each interface (Equations (4) and (5) of this chapter) for the model of the medium discussed in Section 2 of this chapter.

2. Also derive in detail the Equations (12) and (13) obtaining the coefficients a, b, c, d for the model discussed in Section 2.

3. Utilizing the beam splitting technique of Section 3 of this chapter, obtain the transport equations for u and v going to an approximation higher than that demonstrated in this chapter.

4. Show that the wave equation

$$\frac{d^2y}{dx^2} + Q^2(x)y = 0$$

has an exact solution for a profile Q(x) given by

$$\text{(a)} \quad Q^2(x) = \left\{\frac{1}{4} + (C + D \log x)^{-4}\right\} x^{-2}$$

and that the solution is given by

$$y(z) = \frac{1}{\sqrt{q}} e^{\pm i\int_0^z q\, dx},$$

a JWKB like solution where q(x) is given by

$$q^{-1/2} = (C + D \log x)x^{1/2} .$$

(b) Also prove that for a profile

$$Q^2 = (Cx^{\ell+1} + Dx^{-\ell})^{-4} - \frac{\ell(\ell+1)}{x^2}$$

for $\ell \neq -1/2$, we have a JWKB like solution (which is exact)

$$y(z) = \frac{1}{\sqrt{q_1}} e^{\pm i \int^z q_1},$$

where q_1 is given by

$$q_1^2 = Q^2 + \frac{\ell(\ell+1)}{x^2} .$$

Prove this by the method of Chapter II.

CHAPTER VI

1. Start with the Riccati equation given below (in (a)) corresponding to a second order equation, $(d^2u/dx^2) + p^2u = 0$.

(a) $y' + y^2 + p^2(x) = 0.$

Effect change of variables, $y = pv$, and $z = \int p\, dx$. When v is taken constant, we obtain two solutions for y show that corresponding two approximations obtained as

$$u_1 = \exp \int^x q\, dx' \quad \text{and} \quad u_2 = u_2 \int_0^x u_1^{-2}\, ds$$

are solutions of the second order equation. They do not have the turning point difficulty, in some cases since q is given by the approximation

$$q = \frac{-p'}{2p}\left[1 - \left\{1 - \frac{4p^4}{p'^2}\right\}^{1/2}\right]; \qquad p' = \frac{dp}{dx}$$

when p' is large compared with p. This approximation becomes Rayleigh approximation or the WKB approximations when $p'/2p < < 1$, and $y_{1,2} = -(p'/2p) \pm ip$

2. V.A. Bailey in Physical Review 96 (1954), 855 (see Reference [10]) studies the reflection function for non-homogeneous medium, extending between x = a and x = c with momentum p_A for regions below and up to a and p_C and c and beyond. He derives the reflection functions from such a medium with momentum discontinuities at x = a and x = c, as

$$\rho = \frac{\{ip_A - y_0(a)\}f - 1}{\{ip_A + y_0(a)\}f - i},$$

where

$$f = \int_c^a \exp\left(-2\int_a^x y_0\, dx'\right) dx + \frac{\exp\left(-2\int_c^a y_0\, dx'\right)}{ip_C - y_0(c)},$$

where $y_0 = u'/u$ corresponding to one particular solution u of the second order equation. If p and p' are continuous at x = c then the first approximation for ρ

$$\rho_1 = \frac{iP_A - y_0(a)}{ip_A + y_0(s)}$$ where $y_c(a)$ is to be substituted by the

approximation q(a) given in earlier question.

Prove the above statements.

3. In the example of Exercise (2) if p(x) is continuous at x = a, and x = c, show that ρ satisfies the equations derived for the reflection function in Chapter V

$$\rho' = \frac{p'}{2p} - 2ip\rho - \frac{p'}{2p}\rho^2$$

with ρ(c) = 0 and obtain the iterative solutions by quasi-linearization methods. See also Miller, S.O., Jr. and R.H. Good, Jr., Physics Review 91 (1953), 174.

The equation for ρ was also derived by L.R. Walker and N. Wax in Journal of Applied Physics 17 (1946), 1043 [12].

4. In exercise (1) if we substitute the profile $p^2 = (n^2/n^2)$ with $a = n/p_A$ and c = ∞, show that q = β, where

$$\beta = \frac{1}{2n} - \left(\frac{1}{4n^2} - 1\right)^{1/2}$$

and

$$\rho = \frac{i - \beta}{i + \beta} = \frac{i}{2n + (4n^2-1)^{1/2}}$$

5. (i) Convert the second order equations

$$\frac{d^2\varphi}{dx^2} + \lambda^2 k^2 \varphi = 0$$

into a first order equation for the vector $u = \begin{pmatrix}\varphi \\ ik\Psi\end{pmatrix}$ where $\frac{d\varphi}{dx} = ik\Psi$.

(ii) Use the transformation $\vec{u} = \underset{\sim}{P}_0 \vec{v}$ where

$$P_0 = \begin{bmatrix} 1 & 1 \\ \alpha & \beta \end{bmatrix}$$

and obtain the transport equation for $\vec{v}$.

If $\alpha = (g_1'/g_1)$ and $\beta = (g_2'/g_2)$, using the formula given in Chapter V (Section 3), find the coefficients of the transport equation for $\vec{v}$ namely the entities A, B, C, and D in

$$\frac{d\vec{v}}{dx} = \begin{bmatrix} A & B \\ -C & -D \end{bmatrix} \vec{v} .$$

Show that

$$A = \frac{g_1'}{g_1} + \frac{g_2 L(g_1)}{w(g_1, g_2)}$$

$$B = \frac{g_1}{w(g_1, g_2)} L(g_2)$$

$$-C = \frac{g_2}{W(g_2, g_1)} L(g_1)$$

$$-D = \frac{g_2'}{g_2} + \frac{g_1}{W(g_2, g_1)} L(g_2) ,$$

where W is the Wronskian $W = g_1 g_2' - g_2 g_1' = -W(g_2, g_1)$ and L is the same differential operator as in the second order equation we started with.

Make approximately two linearly independent choices of g_1 and g_2 for the equation $L(g) = 0$, say g_1^0 and g_2^0. Ignore the cross-coupling or non-diagonal terms in the transport equation and obtain the solutions $\hat{v}_1^0$ and $\hat{v}_2^0$, Treat these as g_1^1 and g_2^1 for the next iteration and so on. Show that for the higher order solutions

$$g_{1,2}^{n+1} = g_{1,2}^{n} \exp \int^{z} c_{11}^{n}(z)\, dz ,$$

where

$$c_{11}^{n} = \frac{L(g_1^n)g_2^n}{W(g_1^n, g_2^n)}.$$

Hence, show that the general solution φ_1 can be written as an infinite product

$$\varphi_1 = g_1^0 \prod_{n=0}^{\infty} \exp \int^{c} c_{11}^{n}(z)\, dz = \lim_{n\to\infty} g_1^n .$$

(See the article: Bahar, E., Journal of Mathematical Physics 8 (1967), 1735 [4].)

6. Since $c_{11}^0 = [A_0 - (g_1^{0'}/g_1^0)]$, show that the next transformation $\underset{\sim}{P}_1$ such that $\vec{w} = \underset{\sim}{P}_1\vec{v}$, which leads to a higher approximation and $\underset{\sim}{P}_1$ is given by

$$\underset{\sim}{P}_1 = \begin{pmatrix} 1 & 1 \\ A_0 & D_0 \end{pmatrix},$$

where A_0 and D_0 are the transport coefficients obtained after the first transformation. Similarly, the successive transformations are obtained by the transport coefficients obtained by the previous splitting. Prove this statement.

7. (a). If the momentum profile is given by $k^2 = q_0^2 + t(x)$ taking $g_{1,2}^0 \sim \exp \mp i\lambda q_0 x$ as initial approximation when $\|t/q_0^2\| \ll 1$ obtain the usual WKB solution

$$\varphi_{1,2}^1 \sim q^{-1/2} \exp \mp i\lambda \int^{x} q\, dz.$$ What is q?

(b) When gradient of the momentum is slowly varying then represent

$$k^2 = c^2 + ax + r(x); \quad a = \frac{dk^2}{dx}$$

when $\|r'/a\| \ll 1$ and $\|r\| \ll k^2$. Assume the local wave functions as the solution of

$$g''_{1,2} + \lambda^2(c^2 + ax)g_{1,2} = 0$$

$$g_1 = Ai(\xi); \quad g_2 = Bi(\xi)$$

$$\xi = -(\lambda/a)^{2/3}(ax + c^2)$$

and Ai and Bi are the Airy solutions. Compare this with Langer's solutions near the turning point. (R.E. Langer, Physics Review 51 (1937), 669 [11].)

CHAPTER VII

1. We are given the system of equations

$$\frac{\partial u}{\partial x} + \frac{1}{c}\frac{\partial u}{\partial t} = a(z, t)u(z, t) + b(z, t)v(z, t)$$

$$-\frac{\partial v}{\partial x} + \frac{1}{c}\frac{\partial v}{\partial t} = c(z, t)u(z, t) + d(z, t)v(z, t)$$

with $u(z, 0) = v(z, 0) = 0$, $u(0, t) = 0$; $v(x, t) = \delta(t-\tau)$ and a, b, etc., are reasonably arbitrary and well-behaved functions. By considering this generalized transport problem, obtain the corresponding emergent functions by particle counting.

2. Develop a time dependent theory for the many-state case using both classical and imbedding approaches (see Ramakrishnan et al., Reference [9]).

3. In the text the source-free transport problem has been considered. Study the problem with sources S^+ and S^- that are dependent on both z and t. (Use Laplace transform techniques.)

4. Apply direct invariant imbedding theory to diffusion in a rod, starting with Fick's Law. Arrive at the emerging current without using transform techniques (Bellman et al., Journal of Mathematics and Mechanics 9 (1960), 933, Neutron Transport Theory V).

5. Consider the equation

$$y'' + \lambda\sigma^2 y = 0$$

with boundary conditions

$$\beta_1 y'(0) - \sigma y(0) = 0$$

$$y'(x) - \sigma\beta_2 y(x) = 0.$$

Compute the characteristic lengths x for $\beta_1 = \beta_2 = .2$ and $\sigma = 10$, for non-negative values of y [7].

6. Starting with the inhomogeneous time dependent wave equation

$$\varphi_{xx} - c^{-2}(x)\varphi_{tt} = f(x, t) \quad \text{with} \quad \varphi(x, 0) = 0$$

$$\text{and } \varphi_t(x, 0) = 0$$

obtain the coupled set of integral equations for the proper splitting.

CHAPTER VIII

1. For the ℓ - th partial wave Schrödinger equation obtain for the tangent of the phase shift, for a spherically symmetric potentials by the phase approach of Section (3)

$$t_\ell^{(n+1)\prime}(r) = -k^{-1}gv(r)[\hat{j}_\ell^2(kr) - 2\hat{j}_\ell(kr)\hat{n}_\ell(kr)t_\ell^{(n)}(r) + \\ + \hat{n}_\ell^2\, t_\ell^{(n)2}] ,$$

n is the order of approximation. $t_\ell^{(0)} = 0$ and $t_\ell^{(n)}(0) = 0$; g, a constant, is the coupling constant of the potential. The Born phase shift is the lowest approximation and yields

$$(\tan \delta_\ell)_B = -k^{-1}g \int_0^\infty dr\ v(r)\ \hat{j}_\ell^2(kr) .$$

Using quasilinearization obtain the first improved Born approximation

$$(\tan \delta\ell)_{BB} = -k^{-1}g \int dr\ v(r)\hat{j}_\ell^2(kr)\ \exp \{2k^{-1}g \int_r^\infty ds\ v(s)\hat{j}_\ell(ks)\hat{n}_\ell(ks)\}$$

$\hat{j}_\ell(kr)$ and $\hat{n}_\ell(kr)$ are the Riccati-Bessel functions. [See Reference [4]; also Glauber R.G. 'High Energy Collision Th Theory in "Lectures in Theoretical Physics"' (ed. by W.E. Brittin and L.G. Dunham), Wiley, N.Y., 1959.]

2. Study the Schrödinger equation given below for the case $\ell = 0$, relating to scattering from a nonlocal potential

$$u''(r) + k^2 u = \int_0^\infty v(r, s)u(s)\, ds .$$

Arrive at the differential equation for the phase function given by

$$\delta'(r) = -k^{-1} \sin (kr + \delta) \int_0^r ds\, v(r, s) \sin [ks + \delta(s)] \times$$
$$\exp \int_s^r dt\, \delta'(t) \cot [kt + \delta(t)], \text{ with } \delta(0) = 0$$

Consult Reference [4].

In the special case of a sepearble potential $v(r, s) = f(r)g(s)$ the phase function is given by

$$\delta(r) = - \frac{\tan^{-1} \int_0^r ds\, f(s) \sin ks}{\left[A + \int_0^r ds\, f(s) \cos ks\right]}$$

with

$$A = \frac{\left[k - \int_0^\infty dr\, g(r) \int_0^r ds\, f(s) \sin k(r-s)\right]}{\int_0^\infty dr\, g(r) \sin kr}$$

prove the above result.

3. Consider the two equations derived in the text:

$$\Phi = \frac{\sin kr}{k} + \frac{\gamma}{k} \int_0^r \sin k(r-r')\ v(r')\Phi(r')\ dr'$$

$$f(k) = 1 + \gamma \int_0^\infty dr\ e^{ikr}\ v(r)\Phi(r)\ ,$$

carry out the imbedding analysis of these two integral equations in the coupling constant γ. Arrive at the result that $f(k)$ is the Fredholm determinant corresponding to the kernel of the Lippman-Schwinger equation [16] which is given by

$$k(r, r') = -\frac{\sin kr_<}{k}\ e^{ikr_>}\ v(r')\ .$$

($r_<$ and $r_>$ are the usual notations).

4. Using the method of Section 2 study the differential equation

$$\frac{d^2u}{dt^2} + u = f(t)u$$

$$u(0) = d_1 = 0; \qquad u'(0) = d_2 = 1$$

and

$$f(t) = 2 \quad \text{for} \quad 0 \le t \le t_0.$$

Obtain the phase shift and compare it with the analytical solution when t_0 is extended to ∞.

5. Consider the equation

$$y'' - (1+ze^{-at})y = 0; \quad y(0) = c_1; \quad y'(0) = c_2 \quad \text{with} \quad a > 0$$

Set $f(c_1, c_2; z) = \lim_{t\to\infty} y(t)e^{-t}$.

Obtain the partial differential equation satisfied by f.

6. To study an eigenvalue problem by the methods of Section 3 consider the equation

$$\frac{d^2y}{dt^2} - \lambda^2 y(t) = f(t)y(t); \quad 0 \leq t \leq \infty$$

$$y(0) = d_1; \; y'(0) = d_2 \; ;$$

λ is a real parameter. $f(t)$ satisfies the required conditions as given in Section (3). In many physical problems we are interested in those values of λ (if any) such that $y(t) \to 0$ as $t \to \infty$. Analyse this problem by the imbedding technique as was done in the phase shift problem. Consult Chapter 9, Reference [5a] and also Reference [17].

CHAPTER IX

1. Derive the Jacobi Identity for three matrices A, B, C starting with the relation $e^A e^B = e^{A+B+f(A, B)}$ where $f(A, B) = [A, B]/2 + g(A, B) + h(A, B)$. g is a homogeneous polynomial of degree 3 in A and B satisfying g(A, B) and h(A, B) is a sum of homogenous polynomials in A and B beginning with one of degree 4. Use also the associative property of matrices $e^A(e^B e^C) = (c^A e^B)e^C$. See R. Bellman: American Mathematical Society LXV (1958), 180.

2. Show the solution of the equation

$$\frac{dx}{dt} = A(t)x \qquad x(0) = 1$$

is non-singular in any interval $0 \leq t \leq t_1$, in which $\int_0^{t_1} \|A(t)\| \, dt$ exists.

3. The lowest dimensional Lie algebra of physical interest has two elements x and y satisfying $[x, y] = y$. Let $Z = \alpha x + \beta y$. Let $G = ax + by$. Find $G(x) = e^{xz} G e^{-xz}$ in terms of operators x and y and constants a, b, α and β. Hint: G(x) satisfies the differential equation $G'(x) = [z, G(x)]$ $G(0) = G$. See R. Sack Phil. Mag. 3 (1958), 497. 'Taylor Theorem for Shift Operators'. [ans: $a(x) = a$

$$b(x) = \frac{\beta a}{\alpha} + \left(b - \frac{\beta a}{\alpha}\right) e^{\alpha x}]$$

4. Consider the four-dimensional algebra of elements w, x, y, I, satisfying [w, x] = -Sx; [w, y] = Sy; [x, y] = SI. Let $Z = \gamma w + \delta x + \rho y$; $G = gw + dx + ry$. Find $G(x) = e^{xz} G e^{-xz}$, as before. The realization of this algebra is provided by the operators [$a^+ a$, a, a^+ and I], where a and a^+ are the usual annihilation and creation operators occurring in harmonic oscillator problems. See Reference [1].

5. Find the explicit solutions of $u_{xx} + u_{yy} + e^{2kxy} u = 0$. See Reference [10].

6. The representations of the generalized Clifford algebra provide us with (2n+1) matrices, $L_1, L_2, \cdots, L_{2n+1}$ of dimension m^n obeying the generalized Clifford condition

$$L_i L_j = \omega L_j L_i$$

$$L_i^m = 1 ,$$

where ω is the m-th root of unity. For any m, taking the basic representation $L^m = 1$, obtain matrices that obey the idempotent condition

$$L^m = L$$

See Ramakrishnan, <u>L-Matrix Theory or Grammar of Dirac Matrices</u> (Tata-McGraw Hill Publ. Co. Ltd., Bombay 1972).

7. Employing the G.C.A. matrices for which m = 5, and n = 1, obtain the nature of the first W.K.B. type solutions of the equation

$$\varphi^V + \lambda^5 k^5 \varphi = 0 ,$$

where φ^V is the fifth order derivative of φ with respect to the independent variable.

CHAPTER X

1. Consider the equation

$$u'' + [1 + \varepsilon\varphi(t)]u = 0; \quad u(0) = 1; \quad u'(0) = 0.$$

Let u = cos (1 + b_1t(t) and determine b_1, so that

$$J(b_1) = \int_0^{2\pi} (u'' + (1 + \varepsilon\varphi)u)^2 \, dt$$

ia a minimum. Here ε is a small quantity.

2. Prove that the solution of

$$u'' + u + \varepsilon u^3 = 0; \quad u(0) = 1; \quad \text{and} \quad u'(0) = 0$$

is periodic.

3. Consider the equation

$$u' = -u + u^2; \qquad u(0) = c$$

set $u = ae^{-t} + (c-a)e^{-2t}$ and determine a so as to minimize

$$g(a) = \int_0^{\infty} (u' + u - u^2)^2 \, dt.$$

4. Let J(x, y) = (x, Bx) + 2(x, Ay) + (y, By) - 2(a, x) + 2(b, y) where B > 0 and A is not necessarily symmetric. Show that the variational equations Bx + Ay = a; A'x + By = b, are obtained when we minimize over both x and y. Under what conditions does an absolute minimum exist? (B > 0 means B is a politive definite matrix.)

5. Consider the Emden-Fowler equation

$$u'' - x^{-1/2}u^{3/2} = 0; \quad u(0) = 1; \quad u(\infty) = 0.$$

Show that this is the Euler equation of

$$J(u) = \int_0^\infty [u'^2 + \frac{4}{5} x^{-1/2} u^{5/2}]\ dx ,$$

and for the minimizing function u, we have

$$-u'(0) = \int_0^\infty [u'^2 + x^{-1/2} u^{5/2}]\ dx .$$

6. Use the trial function $u(x) = (1 + x^{1/\alpha})^{-\beta}$ and show by appropriate choice of α and β that $\min_u J(u) \le 1.5883$.

J. Ikebe and T. Kato, 'Application of Variational Method to the Thomas Fermi Equations', J. Phys. Soc. Japan 12 (1957), 201-203.)

7. Consider the problem of minimizing

$$J_\lambda(u) = \int_0^T (u'^2 + u^2)\ dt + \lambda(\int u\ dt - a)^2$$

over all admissible u(t) subject to u(0) = c where $\lambda \ge 0$. What is the limiting behavior of the minimum value J(u) and the minimizing function as $\lambda \to \infty$?

8. Consider the problem of minimizing

$$J(u) = \int_0^T u'^2\ dt \quad \text{subject to} \quad \int_0^T u^2\ dt = k,$$

k > 0 with u(0) = c. Consider the associated functional

$$J(u) = \int_0^T (u'^2 + \lambda u^2)\ dt ,$$

where u(0) is again equal to c, $\lambda \ge 0$. Show explicitly that $\int_0^T u^2\ dt$ decreases as λ increases.

9. Consider the integral equation

$$\lambda f(t) = \int_0^a k(t-s)f(s)\, ds,$$

where we suppose k(t) is non-negative, even and monotone decreasing for $0 \leq t \leq \infty$ and that $\int_0^\infty k\, dt < \infty$. Then

$$\lambda_M \geq \frac{2}{a}\int_0^a \left[\int_0^t k(t_1)\, dt_1\right] dt \geq 2\int_0^a k(t)\, dt - \frac{2}{a}\int_0^a tk(t)\, dt .$$

(Bellman, R. and R. Latter, 'On the Integral Equation $\lambda f(x) = \int_0^a k(x-y)f(y)\, dy$, Proceedings of the American Mathematical Society 3 (1952), 884.

10. Let $A_N = (a_{n-m})n$, m = 1, 2, ..., N be a Topelitz matrix and consider the minimization of $[(x, A_N x) - 2(b, x)]$ using dynamic programming. Does this procedure have any advantages over the usual procedures solving $A_N x = b$?

11. Consider the problem of minimizing the function $\sum_{n=0}^{N} [(u_n) + (v_n)]$ over $\{v_n\}$ with $u_{n+1} = au_n + v_n$ and $u_0 = c$. Let $f_N(c)$ denote the minimum value. Show that $f_N(c) = r_N|c|$, where $r_N = \min_v[1 + (v) + r_{N-1}(a + v)]$. From this obtain an explicit relation for r_N connecting r_{N-1}.

12. Treat the problem of minimizing the quadratic expression

$$Q(x) = \sum_{k=1}^{N} c_k(x_k - x_{k-1})^2 + \sum_{k=1}^{N} d_k(x_k - a_k)^2$$

over the quantities $x_1, x_2, \ldots, x_N$ where x_0 is a given constant and c_k and d_k are non-negative constants by dynamic programming methods and obtain the necessary recursion relations.

CHAPTER XI

1. Let v(t) satisfy a Riccati equation $v' + v^2 + pv + q = 0$ and possess a power series expansion

$v(t) = \Sigma_{n=0}^{\infty} a_n t^n$ and p and q are analytic in t. Show how to obtain a continued fraction approximation for v

$$v = a_0 + \cfrac{a_1 t^{c_1}}{1 + \cfrac{b_1 t^{c_2}}{1 + \cdots}}$$

where c_1, c_2, etc., are positive integers.

2. Obtain the solution of the equation $u' = e^{bu} + g(t)$, $u(0) = c$ employing the method of quasilinearization.

3. Consider the equation $r + r^n = a$, $n > 1$; $a > 0$. Starting with the expression $r^n = \max_{S \geq 0} [nrS^{n-1} - (n-1)S^n]$ show that

$$r^n = \min_S \left[\frac{a+(n-1)S^n}{1+nS^{n-1}}\right] .$$

4. Starting with the equation $x''(t) - Bx(t) = 0$, $0 \leq t < T$ with boundary conditions $x(0) = c$; $x'(T) = 0$, arrive at the representation of the square root of the positive definite matrix B, as

$$(c, B^{1/2}c) = \lim_{T\to\infty} \min \left\{ \int_0^T [(x', x') + (x, Bx)]\, dt \right\}.$$

(See Bellman, Introduction to Matrix Analysis, McGraw-Hill, 2nd Edition, 1970.)

5. Using the foregoing representation show that $B_1 > B_2 > 0$ implies that $B_1^{1/2} > B_2^{1/2}$.

6. Consider the equation $u'' - u - 2u^3 = 0$, $u(0) = 0$, $u'(0) = c_1 > 0$, assuming $u(T) = 0$, consider the integral $\int_0^T u(u'' - u - 2u^3)\, dt$ and show that u is non-zero for $t > 0$.

CHAPTER XII

1. Derive the cardinal spline by dynamic programming methods.

2. Solve the integral equation

$$y = 1 + \int_0^1 k(x, t)y(t)\, dt$$

by the method of splines with $\alpha^2 = 100$

$$\begin{aligned} k(x, t) &= \alpha^2(x-1)t \quad (0 \le t \le x) \\ &= \alpha^2 x(t-1) \quad (x \le t \le 1) \end{aligned}$$

Compare the numerical values with exact solution.

3. Derive the recursive relations for the fifth-order spline approximation given the values $y_1, y_2, \ldots, y_N$ at N points of the curve $y = f(x)$ at $x_1 \ldots x_N$. (Use the methods of this chapter.)

4. Solve the following equation numerically by differential quadrature

$$u_t(x, t) = x^2 - \frac{1}{4}[u_x^2(x, t)]$$

with $u(x, 0) = 0$ and $u(0, t) = 0$. Compare it with the actual solution $u(x, t) = x^2 \tanh t$.

INDEX OF NAMES

INDEX OF SUBJECTS

GPSR Compliance

The European Union's (EU) General Product Safety Regulation (GPSR) is a set of rules that requires consumer products to be safe and our obligations to ensure this.

If you have any concerns about our products, you can contact us on ProductSafety@springernature.com

In case Publisher is established outside the EU, the EU authorized representative is:

Springer Nature Customer Service Center GmbH
Europaplatz 3
69115 Heidelberg, Germany

Batch number: 10607212

Printed by Printforce, the Netherlands